Fast-Tracking Your Career

Fast-Tracking Your Career

SOFT SKILLS FOR ENGINEERING AND IT PROFESSIONALS

WUSHOW "BILL" CHOU

Hard Skills Help Us Qualify for a Job;
Soft Skills Dictate Our Career Growth

IEEE PRESS

WILEY

Published by John Wiley & Sons, Inc., Hoboken, New Jersey.
Published simultaneously in Canada.

For general information on our other products and services or for technical support, please contact our Customer Care Department within the United States at (800) 762-2974, outside the United States at (317) 572-3993 or fax (317) 572-4002.

Wiley also publishes its books in a variety of electronic formats. Some content that appears in print may not be available in electronic formats. For more information about Wiley products, visit our web site at www.wiley.com.

Library of Congress Cataloging-in-Publication Data:

Chou, Wushow.
   Fast-tracking your career : Soft Skills for Engineering and IT Professionals / Wushow "Bill" Chou.
      pages cm
   ISBN 978-1-118-52178-6 (pbk.)
   1. Engineering–Vocational guidance.   2. Information technology–Vocational guidance.
3. Soft skills.   I. Title.
   TA157.C477 2013
   620.0023–dc23

2013002854

Printed in the United States of America

10 9 8 7 6 5 4 3 2

Fundamental Soft Skill Principle:
CULTIVATING GAIN-GAIN PERSPECTIVES

The fundamental principle behind any soft skill is to cultivate the perception in other people's minds that they can gain and benefit by engaging with us.

(Author's special note: Many examples used in the book to illustrate this fundamental principle are based on people in high positions, such as CEOs, CIOs, and VPs, and/or on people working in the engineering and IT fields. However, the principle behind these examples is equally applicable to any position and to any professional field.)

DEDICATION

This book is a collection of my observations on the importance of soft skills throughout my career. I dedicate this book to my wife, Lena, as recognition for her support and encouragement at various stages during my career.

Contents

PART TWO

Dealing with People: The Essential

PART THREE

Dealing with the Self: The Basic

PART SIX

Being Visionary: Leading to the C-Suite

Foreword

Dr. Sorel Reisman, *Past President (2011), IEEE Computer Society*

It seems like only yesterday that I graduated from the University of Toronto and started my first job as a new electrical engineer at Canadian General Electric. If I had only known then what I know now, my professional career might have been smoother and evolved in entirely different ways. It's not that my career has been bad—it hasn't. But life would have been much easier had I been aware and prepared for some of the critical decision points described in this book before I encountered them. For example, in retrospect and almost by chance, I did inadvertently follow Bill's advice regarding Communications (Chapter 1). Early in my career at IBM, I learned the importance of being "Communications Smart," with the subsequent benefit of being noticed by my managers and being sent to a variety of management training programs—probably the most useful sessions I've attended in my career.

When Bill asked me to review an early draft of this book, I was in the middle of extensive business travel and was reluctant to take on the task of reading the draft. However, shortly after I started the book, I couldn't put it down. That draft traveled with me on many airplanes around the world, and kept me entertained, reminiscing about situations in my career that were so similar to many described by Bill. In fact, time after time, I stopped my reading and interrupted my wife to say, "Listen to this." Then I would read her one of the vignettes from the book and ask her if she remembered when I had encountered the same situation at company XYZ. Inevitably, she would ask me, "So what is Bill's advice about this situation?" And inevitably after I told her, she would say, "Too bad you hadn't read Bill's book or had Bill with you at that time. Your life would've been a lot easier."

The brilliance of this book is how accurately Bill has described so many critical, career-altering situations that every professional encounters in their life. And the rewards for readers, young and not-so-young are significant. This book can serve as an instruction manual and mentor for at every stage of a career. For example, readers who are early in their career, who have yet to encounter situations described in the book, will be on notice and better prepared for circumstances they will likely encounter down the road. Those who are towards the end of their career will, as I was, be entertained as they go down memory lane. And although the cases in the book concern technology industries, and while I have spent my life in the computer industry, I think that the content applies to any industry.

Much of the content of this book will resonate with and be appreciated by career professionals who, as I have, lived through them. One can only hope that young career professionals will read this book, internalize Bill's advice, and be better prepared when they inevitably encounter the same career issues that we all do.

From a personal standpoint, knowing Bill for almost 20 years, I have gained a much better appreciation of "encounters" he and I have had, where his wisdom inevitably prevailed. I can see now how Bill's sensitivity to complex professional situations and environments has allowed him to be so successful in his own career. All of us, as professionals, should be thankful that Bill has taken the time to produce this wonderful work that can help new and mid-level professionals achieve the goals to which they aspire—and also to entertain us "old-timers."

I thank you, Bill, for giving me the opportunity to participate in the production of your work and with it, to reminisce about my own career successes and failures.

Guest Introduction I

Dr. Simon Y. Liu, *Editor in Chief, IT Professional, 2010–2013*

I am delighted to see this book on soft skills for engineering and IT professionals by Dr. Chou. This book covers essential skills for career planning, development, and advancement. Engineering and IT professionals often overly focus on hard skills for performing tasks. They need practical and useful guidance on soft skills to enhance communications, interpersonal interactions, job performance, and career advancement. Unfortunately, there are relatively few books that offer such guidance, especially for engineering and IT professionals.

Dr. Chou is an accomplished IT leader, a distinguished educator, and a prolific writer. His connections and involvement with government leaders, business executives, brilliant engineers, IT professionals, and IT consumers add a tremendous amount of real-world insight and relevance. His personal experiences as both practitioner and educator are also clear throughout the book. I had to learn many of the soft skills covered in this book though real-world experience with my fair share of mistakes.

This book is an excellent resource that will give you the necessary knowledge and tools to put your career on the fast track. The book is a one-stop shop that clarifies a variety of mystical topics in career development. If the mistakes made and lessons learned through my career journey are any indication, this book will be used every day by engineering and IT professionals interested in continuous and sustainable career development. I congratulate Dr. Bill Chou on this excellent book, which provides an invaluable resource. I found this book exceptionally practical and extremely useful and I believe you will, too. Enjoy!

Guest Introduction II

Dr. Arnold "Jay" Bragg, *Editor in Chief, IT Professional, 2006–2009*

Bill Chou began writing a series of short articles for *IT Professional* magazine (an IEEE Computer Society technical publication) during my tenure as Editor in Chief. We ran the articles as part of the magazine's Developing Soft Skills department. Bill's series was very popular with our readership, and comments were overwhelmingly positive. One reader suggested sending copies of Bill's articles to Dilbert, thinking they might help. Another reader, a 43-year veteran of the IT trenches who had climbed the ladder from keypunch operator to CIO, said, "Chou gets it. He really [expletive] gets it. He must be famous."

Famous indeed! Bill is Professor Emeritus of Computer Science; an IEEE Fellow; the first CIO at the US Treasury Department; a successful consultant, entrepreneur, author, and editor; an engaging and entertaining speaker; and a distinguished researcher in telecommunications and computer networking. I've known Bill for more than 30 years, and consider him to be the ultimate high achiever.

Bill also "really gets it." He is a co-founder of *IT Pro* and served as the magazine's first Editor in Chief. We've run hundreds of articles, tutorials, and case studies in the past 15 years—many written by distinguished academics—and have always made sure that each was both appealing to IT practitioners and true to Bill's vision.

Bill's book, *Fast-Tracking Your Career: Soft Skills for Engineering and IT Professionals*, was inspired by his Developing Soft Skills series in *IT Pro* and guided by lessons learned during his distinguished career in IT. If you Google® "soft skills for IT professionals," you'll get more than 3.5 million hits. However, there are few books on the list, and many of the articles and presentations focus solely on the importance of communication skills, how to make effective presentations, and how to build professional relationships. Bill's book is much more than that.

An important differentiator is the "fast-tracking your career" thread in every chapter. Soft skills are critically important to engineering and technical professionals, who often downplay the "soft" side. Bill explains which skills arc really important, and why. Unless you're managerially brilliant—and few of us are—success in your project, in your job, in your career, in your life, depends on how effectively you manage your time, your relationships, your managers and staff, and your decision making processes.

I love pithy laws/rules/conjectures/propositions/principles, and use them in meetings when I'm losing an argument. I quote Brooks' Law—"Adding manpower to a late [software] project makes it later"—at least once a month, and no one ever disagrees. I've found the Markovnikov Rule—"Them that has, gets"—and its negation, the Anti-Markovnikov Rule, to be 100% effective with senior management unless they happen to be organic chemists. I've used Wirth's Law—"Software gets slower faster than hardware gets faster"—to justify six-figure hardware upgrades. I've quoted Letwin's Conjecture—"One in a million is next Tuesday"—to get another day/week/month/quarter for testing added to a schedule. I've used Post's Proposition—"[A] real programmer can write FORTRAN programs in any language"—in arguments with Real Programmers unless they've never heard of FORTRAN, in which case I substitute the Real Programmer's favorite language. Et cetera.

Bill's book is full of examples in which soft skills make a difference. Each can be reduced to one or two key principles. My favorite, which I quote at least once a week, is "don't polish a sneaker." The idea is simple: sometimes we must quietly scream "Enough!," stop what we've been doing, and take a new approach. Bill uses half a dozen examples to illustrate how to know that it's time to stop polishing that sneaker: to give up on projects or tasks that are unlikely to progress as we had intended; to move on to another organization or position if the boss is inept or unreasonable or if we are unlikely to advance as we had hoped; to give up on people who are unlikely to progress as we had expected; or to give up on visions that are unlikely to unfold as we had dreamed.

And knowing when to stop polishing that sneaker is indeed a soft skill. It is related more to human nature than technical prowess, and if wisely applied, it will almost certainly contribute to one's career growth. The best career advice I ever received came from a friend from my father's generation, a gruff old statistician at Purdue University. Rod asked me about my projects, boss, staff, and visions, listened patiently, and then said, "Don't you see? Finish up and get out!" Hey, that's the "don't polish a sneaker" principle.

I'm sure you'll find *Fast-Tracking Your Career: Soft Skills for Engineering and IT Professionals* an interesting read.

Guest Introduction III

Frank E. Ferrante, *Editor-in-Chief, IT Professional,*
2002–2005

Dr. Wushow (Bill) Chou, for many years a technical giant in the innovation of new packet network architecture optimization protocols that supported the foundation of our current global Internet systems, a consultant, a teacher, and a friend, has been an astute observer of events and decisions that successfully propelled him and others upward in their careers. In this book, Dr. Chou offers the perspective of one who is intimately familiar with the successful choices one needs to make if he or she is to reach his or her professional career goals. Many who recognized his leadership in his professional career respect Dr. Chou for his insight into the importance of how careful one should be in responding to opportunities that affect career changes.

From his vantage point as one whose career at times appeared to be limited due to his language barrier or his lack of time to focus on his many projects, Chou's decisions appeared to almost naturally surge forward at light speed as he applied his analytic training to the opportunities he determined were worth his time and attention to detail. Following his successful leadership as the founding Editor in Chief of the IEEE Computer Society's *IT Professional* magazine (*IT Pro*) in 1999, Chou not only demonstrated his prowess at making his publications a success, but over time it has been shown that his decisions early in its establishment laid the groundwork for future Editors in Chief that has resulted in the publication's continued successful operation. Finally, as he tried to retire from his front-line editorial role, Chou, now named the chair of the Editorial Board of Advisors for *IT Pro*, took time to offer his perspective on career fast-tracking in a series of short yet prophetic articles. His articles were aimed at helping others to gain insight into what he had been carefully compiling from his career's observations. He noted choices he and his compatriots have been making over the years that supported a rapid acceleration upward in their careers. Articles published in *IT Pro*'s pages such as "Communication Smart" (March/April 2009), "People Smart" (May/June 2009), and "Job Interview Smart" (July/August 2010) started to open up his knowledge base in the field of soft skills. Chou felt, as he stated in his proposal for writing this book, that "for most individuals, success in one's career depends substantially on one's soft skills (human nature related skills), less on one's hard skills (professional skills)." In varied examples throughout the book, Chou not only describes his own experience in applying soft skills, but he also includes the successes his friends have realized by their actions pertaining to soft skills application (knowingly or unknowingly, i.e., naturally applied). Chou has commented many times that people in the top tier appear to have one thing in common: they are all good in soft skills. What Chou offers you, as reader, is a truly masterful compilation of examples and a most comprehensive identification of details related to soft skill applications. The analytically organized treatise that Chou offers, if mastered, should help fellow technical professionals to advance their careers even further and faster than his own as they choose among through their life's opportunities for advancement. Indeed, this book is a trove of golden rules that you will read, enjoy, and treasure as you work your way through your career and life.

Preface

Based on my 25 years of experience as a senior manager/executive in industry, academia, and government, and in my work as a consultant for over 30 large organizations, I have made the following observations:

1. Success in one's career depends substantially on one's soft skills (human nature related) and less on one's hard skills (profession related).

2. Professionals, particularly engineers and other highly technical people, are woefully lacking, and showing inadequate interests, in soft skills.

In this book, I summarize a set of rules that will help you put your career on the fast track. This material is based on the "Developing Soft Skills" columns I wrote for *IT Professional*, an IEEE Computer Society magazine, during 2009 and 2010.

When I first arrived at UC Berkeley as an international graduate student, I immediately realized that in addition to having much better verbal skills, many of my schoolmates were smarter than I was and had a stronger scholastic background than I did. When I started to work, I found the same was true for many of my colleagues. Common logic would dictate that most of them should have had a more successful professional reputation and career than I have had. But, in fact, that is not so. Why? I think it is because of the following fact of life that I have observed firsthand: the people at the top tier of a profession are not necessarily the most knowledgeable ones in that profession, and they are not even necessarily the smartest. I have further noticed that good verbal skills do not necessarily result in good communication skills, which are far more important than speaking and writing with flair and eloquence. *But people in the top tier do have one thing in common: they all excel in soft skills.*

In this book, I present the soft skills that I have gathered from observing how successful senior managers/executives moved up their career ladders, share the soft skills that have served me well, and share lessons learned from my own, as well others', failures. I wish I had known and mastered these skills when I was still young.

More than 150 examples illustrate these skills. The examples are real events with real people: students, technical staff, first-level managers, middle-level managers, senior managers, senior executives, professors, public figures, and a few others.

Even though the people referred to in these examples are mostly senior engineering professionals—in particular, IT professionals—the basic principles behind the examples and the related soft skills transcend all professional fields and levels. I am very confident that any reader who can master most of the soft skills in this book will find him- or herself on the fast track to a highly successful career.

Wushow "Bill" Chou
Chair of Advisory Board, IT Professional

Acknowledgments

Contributions made by Dr. Arnold "Jay" Bragg, Professor Frank Ferrante, Dr. Andy Lee, Dr. Sorel Reisman, Dr. Simon Liu, and Dr. Linda Wilbanks have substantially enhanced this book from its original draft manuscript. All these individuals have achieved one or more C-level positions in their careers. When I sent them my draft manuscript for review, they gave me many critical, some even biting, yet constructive comments. From their feedbacks, I picked through their combined wisdom and experience and incorporated some of their ideas into the final product, this book. I am indeed indebted to all of them.

I am very grateful to Dr. Howard Frank, Chancellor Larry Montieth, and Honorable George Munoz. They were my bosses and my mentors. I learned a great deal of smart soft skills from them.

Ms. Jennifer Stout, a highly regarded copyeditor, did a wonderful job in editing the book. It is through her effort that the book reads smoothly. I am very appreciative.

About the Author

Wushow "Bill" Chou

PhD in electrical and computer engineering, UC Berkeley

Founding Vice President of Telecommunications, Network Analysis Corp.
Founding Director of Computer Studies, Professor Emeritus, North Carolina State University
Founding Chief Information Officer, Deputy Asstistant Secretary, US Department of Treasury
Founding Editor in Chief, *Journal of Telecommunications*, Computer Science Press
Founding Editor in Chief, Chair of Advisory Board, *IT Professional* magazine, IEEE Computer Society

Oversaw IT operations at 14 bureaus (14 CIOs, 11,000+ staff, and $2B annual budget)
Supervised operation of 4,000-node communication network ($285M annual budget)
Consulted for 30+ organizations (US/international agencies and companies)

Frequent paid professional speaker
100 publications
Chaired four IT conferences
Chaired three federal government-wide IT committees
Managing Editor, *Networks*, John Wiley & Sons

Fellow, IEEE (Institute of Electrical and Electronic Engineering)
Meritorious Service Award/Medal, US Department of Treasury

Introduction and Summary

For a fraction of 1% of people, things always seem to go their way, whether because they are extremely smart or extraordinarily lucky. But for the remaining 99+% of us, our career success depends substantially on the deftness of our soft skills (personal nature related) and the proficiency of our hard skills (profession related).

Even though soft skills often play a critical role in career advancement, many professionals, especially engineers and other highly technical people, pay little attention to this fact.

This book identifies 11 functional soft skills that are crucial for putting us on the career fast track, which is critical for achieving a successful career. *A fast-tracked career leads to a successful career.*

Ten of them are required for working successfully within the "box" of a normal professional environment: Communications Smart, People Smart, Work Smart, Time Smart, Career Smart, Marketing Smart, Job-Interview Smart, Boss Smart, Motivating Smart, and Delegating Smart.

The remaining skill is essential for thinking and working successfully beyond the confines of that "box": *Beyond the Box*. All the senior executives whom I have had opportunities to observe excel in this particular skill. Anyone aspires for the C-Suite should master this skill.

ENGINEERS ARE POTENTIALLY BETTER POSITIONED AS EXECUTIVES

Nowadays, a successful senior executive needs to be astute in both business and technology. The desired strengths of astuteness depend, of course, on the specific responsibility the executive is charged with. (Business astuteness refers to business strategy, marketing, investment, organization, and management. Technology astuteness refers to projection of technology advancement, applying technology to mission enhancement, conversion to newer technology, and software/hardware development.)

Fast-Tracking Your Career: Soft Skills for Engineering and IT Professionals, First Edition.
Wushow "Bill" Chou.
© 2013 The Institute of Electrical and Electronics Engineers, Inc. Published 2013 by John Wiley & Sons, Inc.

Neither business astuteness, nor technology astuteness, is directly taught at school. But it is much easier for an individual with an engineering background to become savvy in employing and developing technology than an individual with only a business background. Engineers are advantaged in the technology aspect.

Furthermore, due to his or her inherent logic training, an engineering professional can also easily and quickly learn to be savvy in running a business if he or she is motivated in doing so. This is evidenced by the fact that many high-tech companies have been successfully run and/or founded by relatively *young* engineering/technical professionals, such as Microsoft, Apple, HP, Yahoo, Google, Oracle, Sun Microsystems, Intel, Lenovo, and Facebook, to name just a few.

One may argue that these companies are exceptional cases. But the ability for young engineers to easily and quickly learn to be savvy in running a business is nothing unusual. I personally know at least four engineers who, highly motivated in moving up the managerial ladder, have taken less than 5 years to advance from technical staff positions to those of senior managers/executives.

Yet here is the big *irony*. While engineers are very cable of learning to be business astute, most do not have the mindset to do so. On the other hand, individuals who do not have an engineering or scientific background are more attuned to learn to be savvy with business. Similarly, while engineers are quite cable of mastering soft skills, most show little interest in doing so. (Some engineers even snub soft skills.) On the other hand, individuals without an engineering or scientific background are more willing to do so.

In short, if an engineer is motivated and aspires to move up to senior executive positions, he/she has a big advantage over nonengineers, particularly in technology-oriented organizations.

CATEGORIZATION OF SMART SOFT SKILLS

From the perspectives of objectives and functionalities, the above-mentioned 11 soft skills may be organized into the following six categories.

Communicating: Communications Smart

The absolutely necessary skill is the ability to communicate successfully.

More specifically, it is the ability to communicate effectively to an audience what we want to convey to them.

Dealing with People: People Smart, Marketing Smart

Beyond smart communication skills, our success depends on our skills in dealing with people.

More specifically, it depends on how respectable and likable we are in getting the right people to work with us and help us when needed and on how persuasive we are in promoting our work and ourselves.

In a strict sense, *Communications Smart* is a component of *People Smart* and *Marketing Smart* is a special case of *Communications Smart*. I single them out with separate chapters to highlight their importance in our career advancement.

Dealing with Self: Work Smart, Time Smart, Career Smart

Beyond the absolutely necessary skill of communicating and the essential skill of dealing with people, our success depends on our basic skills in managing ourselves.

It depends on how effective we are in carrying out our work, on how effective we are in dispensing our time, and on how methodical we are in planning our career.

Dealing with the Boss: Job-Interview Smart, Boss Smart

Equipped with the absolutely necessary, the essentials and the basics, our success now depends substantially on how well we can earn trust and recognition from our bosses.

This process begins at the end of a successful job interview. If we are offered a position, we have likely done well enough in the interview process and have earned from the hiring manager the trust and recognition that we can perform well in the position. Once we start the job, our success now depends on our ability to cultivate and enhance this trust and recognition.

Dealing with the Staff: Motivating Smart, Delegating Smart

If we are a manager with direct reports, how well we perform on tasks charged to us depends substantially on how well our staff performs.

Consequently, the level of our success now hinges on how well we can motivate them and how astutely we can match tasks with our staff's interests and abilities.

Being Visionary: Beyond the Box

What separates a very successful career from being merely a successful one depends on how farsighted we are in forming and implementing a vision. This in turn depends on how clever and creative we are in thinking beyond the box. This is an essential skill in getting into the C-Suite.

Table I.1 summarizes this grouping of the 11 soft skills into the six categories.

RULES FOR MASTERING SMART SOFT SKILLS

In this book, each soft skill is designated with its own chapter, along with some key rules that are needed to improve our deftness with the skill.

Some rules are equally important across several soft skills and appear multiple times so that each chapter stands on its own—for example, "resonance," with some variations, is equally important to "Communication," "Marketing," and "Job Interview."

Some rules, while still important to multiple soft skills, are not listed or explained in all these skills' chapters. One example is "never polishing a sneaker." It is equally important to "Career," "Work," and "Boss," but it is only highlighted in the "Career" chapter. Most people do not need multiple explanations from me for why they should not "polish a sneaker" if they encounter an unworkable task or a real lousy boss.

TABLE I.1   Categorization of Soft Skill Functions

Categories	Functions	Note	
Communicating (the absolutely necessary)	Communications Smart		Soft skills for within the box of a normal working environment
Dealing with people (the essential)	People Smart Marketing Smart	"Marketing Smart" is a special case of "Communications Smart"	
Dealing with self (the basics)	Work Smart Time Smart Career Smart		
Dealing with boss (earning trust and recognition)	Job-Interview Smart Boss Smart	"Job-Interview Smart" is a special case of "Marketing Smart"	
Dealing with staff (inspiring loyalty and productivity)	Motivating Smart Delegating Smart		
Being visionary (leading to the C-suite and other great opportunities)	Beyond the Box		Soft skills for outside the box of a normal working environment

Table I.2 is a matrix that associates rules with soft skills. These associations are identified as follows:

- A symbol "X," marking a rule/soft skill pairing, indicates that the rule is listed and explained in the chapter designated for that skill.
- A symbol "*," marking a rule/soft skill pairing, indicates that the rule is NOT listed in the chapter designated with that skill. The rule is important to that skill, but is listed and explained elsewhere in the book.

For more information and detailed discussions on the rules, please refer to the Appendix, and Chapters 1 through 11.

- *Appendix*
    - *Table A.1:* Summarizes principles and strategies of the 11 soft skills
    - *Table A.2, Table A.3, Table A.4, Table A.5, Table A.6, Table A.7, Table A.8, Table A.9, Table A.10, Table A.11, and Table A.12:* Each table summarizes key aspects of one soft skill with the following information:
        - The basic principle of the soft skill
        - The basic strategy to acquire the soft skill

TABLE I.2 Rules for Mastering Soft Skills

	Communications Smart	People Smart	Marketing Smart	Work Smart	Time Smart	Career Smart	Job-Interview Smart	Boss Smart	Motivating Smart	Delegating Smart	Beyond the Box
Being always ready for elevator pitches/speeches (in order to express well at opportune encounters)	X										
Preparing targeted elevator pitches/speeches (in order to express well at scheduled meetings)							X				
Mastering a presentation by mastering the onset	X										
Using three diagrams to simplify complexity	X										
Sizing up and resonating with the audience	X										
Sizing up and resonating with our "customers"			X								
Sizing up and resonating with the interviewer							X				
Avoiding blunders of overconfidence				X							
Avoiding gaffes by avoiding overconfidence		*					X				
Being careful of careless comments	X		*				*	*			
Using plain language	X		*				*	*			
Getting accepted by accepting others first		X									
Winning by understanding both ourselves and our counterparts		X									
Being aggressive by being nonaggressive		X									

(Continued)

TABLE I.2 (*Continued*)

	Communications Smart	People Smart	Marketing Smart	Work Smart	Time Smart	Career Smart	Job-Interview Smart	Boss Smart	Motivating Smart	Delegating Smart	Beyond the Box
Gaining by giving		X									
Successful networking by networking less		X									
Being heard by listening	*	X									
Livening up presentations by using jokes and self-deprecating humor	X										
Getting liked by displaying self-deprecating humor (in conversations)		*									
Achieving outstanding results by not seeking perfection				X							
Focusing on self-examination, not on blaming others, when things gone awry				X							
Investing time with the same zeal as venture capitalists investing money					X						
Killing two birds with one stone				*	X						
Minding ROI (return on investment)					X						
Making nonproductive time productive					X						
Turning spare time into opportunities					X						
Keeping the mind sharp by taking catnaps					X						
Opting to be a big fish in a small pond						X					
Hopping to a more opportune pond at opportune moments						X					
Never polishing a sneaker				*		X		*	*		*
Making a good lasting impression by making a good first one						X					
Putting a positive spin on our "product"			X								

6

Putting a positive spin on our qualifications					X
Making a convincing presentation with a well-crafted presentation	*				X
Inciting enthusiasm with enthusiasm	*				X
Winning interviewers' confidence in us by exhibiting confidence	X				X
Being well prepared by collecting relevant information	X				X
Winning trust by showing loyalty				X	
Winning loyalty by being loyal			X	X	
Gaining gratitude by sharing credit and taking blame			X	X	
Getting credit by not taking credit			X	X	
Being astute by watching for nuances				X	
Being proactive and farsighted	*			X	
Showing enthusiasm for challenging assignments				X	
Motivating by complimenting			X		
Getting more done by doing less		X			
Delegating successfully by matching tasks with staff		X			
Making controversial decisions by not making them		X			
Examining the big picture to identify opportunities	X				
Forming a visionary plan	X				
Marketing the vision	X				

"X" indicates that the rule is an important one to the soft skill and is defined in that skill's designated chapter.

"*" indicates that the rule is an important one to the soft skill but is not defined in that skill's designated chapter. It is explained in the chapter of the soft skill where "X" is marked.

- ■ The basic rules needed to improve deftness with the soft skill
- ■ Extended definitions of these rules.
- *Chapter 1 through Chapter 11*
    - ○ Each chapter discusses rules associated with one soft skill.

RELATIONSHIPS AMONG THE SOFT SKILLS

Interdependence

The effectiveness of any soft skill depends substantively on its synergy with other soft skills.

To excel with any soft skill function, we cannot limit ourselves just to focusing on rules that are targeted for a particular function. Inevitably, we will have to judiciously incorporate and complement them with rules related to other soft skill functions.

For example, if we need to develop a strong win-win working relationship with our boss to enhance our career future ("Boss Smart"), it is not enough merely to focus on rules that directly apply to "Boss Smart." It would be very helpful, or even necessary, to also be a good communicator ("Communications Smart"), to have a good relationship with people ("People Smart"), and so on. Conversely, if we already have good commands of "Communications Smart," "People Smart," and so on, it would be relatively easy to develop a good command of "Boss Smart."

Figure I.1 depicts this interdependence relationship.

FIGURE I.1   Interdependence among soft skills

Parts making up the whole

Soft skills for career advancement are multidimensional and complement each other. (They can even be symbiotic.) The 11 soft skills, as identified in this book, provide an effective framework for fast-tracking a successful career.

Although the topic of professional skills is not the subject of this book, remember that as we move up the career ladder, the required professional skill set is bound to change. We will need to evolve our skill set in order to have a good command of professional skills befitting whatever positions we are holding. If we are members of a technical staff, we need to have an *in-depth* understanding of the technical knowledge of the projects we are working on. If we are senior technical managers, we need to have a good (but not necessarily in-depth) understanding of a *broad* spectrum of technical/professional issues related to our responsibilities. If we are senior executives, or members of the so-called C-Suite, we need to have a good command of appropriate *business* knowledge and astuteness related to our mission.

Figure I.2 depicts this "parts making up the whole" concept.

FIGURE I.2 Parts making up the whole

PART ONE

Communications
THE ABSOLUTELY NECESSARY

Communications Smart

Principle: Staying succinct and focused

Strategy: Get key points across within the audience's limited attention span

The absolutely necessary soft skill is good communication: the ability to successfully convey to the audience what we want them to know. Those who are successful in their professional field are almost certainly good communicators.

President Ronald Reagan and Dr. Stephen Hawking have one thing in common: being the greatest communicators in their respective fields. President Reagan owed his success in garnering congressional support for his policies to his excellent skill at persuasion. Dr. Hawking, a very popular theoretic physicist, owes his popularity to his ability of explaining to laymen the esoteric concepts of black holes and time–space relationships in an easy-to-understand fashion. (Hawking has achieved this feat in spite of his severe handicaps in speaking and writing.)

Communication is not simply about writing articles or giving speeches. It is the ability to enable listeners and readers to understand what we are trying to convey to them. Accordingly, we must focus on what we want to deliver in a way that our audience can comprehend and appreciate.

Practicing the following rules, defined later in this chapter, can help us achieve deftness in being Communications Smart:

- Being always ready for elevator pitches/speeches
- Mastering a presentation by mastering the onset
- Using three diagrams to simplify complexity

Fast-Tracking Your Career: Soft Skills for Engineering and IT Professionals, First Edition.
Wushow "Bill" Chou.
© 2013 The Institute of Electrical and Electronics Engineers, Inc. Published 2013 by John Wiley & Sons, Inc.

- Sizing up and resonating with the audience
- Being careful of careless comments
- Using plain language
- Using jokes and self-deprecating humor
- Being heard by listening (Refer to Chapter 2, People Smart, for a general discussion of this rule.)
- Making a convincing presentation by making a well-crafted presentation. (Refer to Chapter 3, Marketing Smart, for a general discussion of this rule.)

RULE 1: Being always ready for elevator pitches/speeches

Be prepared to articulate short pitches at any brief opportune encounter in order to make the right impressions with the right people.

We all have had the experience of a chance encounter, such as in an elevator or at a cocktail party, with someone we would like to impress. We may also have had the experience of expected encounters with unscripted questions, such as during a job interview. In such situations, we usually have only a couple of minutes to say what is needed to impress our listeners.

Of course, we can form a response on the spot. Indeed, some people are very talented at launching a good pitch without prior preparation. But for most of us, it is more likely that we can give a better response if we are prepared.

It is therefore prudent for most of us to prepare a set of "elevator pitches" for those situations in which we need to answer questions on the fly. These pitches can also be tweaked as situations change—undoubtedly, some will need to have different versions for a different audience.

If we do encounter the need to give an unprepared short pitch and we are unable to give a good one right away, we can use a three-point response strategy to make up a pitch as shown in an example later in this section.

One important point: when delivering an elevator pitch, it should sound like an ongoing conversation, not a recitation from a prepared written statement.

The elevator pitch as it has been discussed so far is used as part of a conversation or Q&A. There is another type that is used as a response to an expected request for a short speech/presentation to a group of people. For distinguishing the two types, I shall name this variation as "elevator speech." The occurrence of needing an elevator pitch is much more frequent than that of an elevator speech. However, when we encounter an unexpected need to give an elevator speech on the spot, we can be in quite an awkward situation if we are not prepared and do not have anything meaningful to say.

EXAMPLE: A good "elevator pitch" led to a promising career

Vincent worked as a part-time caddy at a golf club when he was a university student. He was frequently assigned to the same elderly gentleman. During a seemingly casual conversation, Vincent's comments about a market trend impressed the old man. The next day, the part-time caddy got a call from a Wall Street company and was offered an internship position. It just so happened that the elderly gentleman was the CEO of the company.

After Vincent graduated, he joined the firm. In the ensuing years, he became a protégée of the CEO, and today, Vincent is a senior VP at another large Wall Street firm.

EXAMPLE: Well-prepared "elevator pitches" led to passing a PhD exam with ease and a prompt promotion

Eddy was working toward his PhD at a university. One process he had to go through was an oral exam with a committee of three faculty members. A typical student would put most of his attention on hitting the books. Eddy took a slightly different approach. From his standpoint, an oral exam consisted of several Q&As, so the best strategy was to prepare a set of "elevator pitches" to potential questions. With this in mind, he sought out those who had been previously examined by anyone sitting on his oral exam committee. He collected a large number of questions that had been asked by these committee members and prepared an "elevator pitch" for each question. (A typical student would also collect potential questions, but not as methodically and thoroughly as Eddy did.)

Sure enough, some of the same questions were asked during Eddy's oral exam, and he was able to answer them with ease. He not only passed the exam, he also impressed the professors on the committee.

One committee member was the department chair. After the exam, he promoted Eddy from the position of teaching assistant to that of teaching fellow.

Another committee member later claimed that of all the oral exams in which he has participated, Eddy performed the best. He has since helped Eddy's career in a variety of ways.

EXAMPLE: Enormous embarrassment for having not prepared an "elevator speech"

The IRS (Internal Revenue Service) has over 10,000 IT professionals and other staff supporting its IT systems, and its IT organization has several divisions. Thus, a division-wide meeting for IT professional staff often needs to be held in an auditorium.

When I was the Deputy Assistant Secretary for Information Systems and Chief Information Officer at the US Department of Treasury, I had the responsibility of overseeing the IT operations of all bureaus, including the

IRS, under the Treasury Department. One day, I got a request from an IT division director at the IRS to present an award to a staff member in his division. I agreed. My plan was simply to walk in at the appointed time, present the award, congratulate the award recipient, and leave right away. When I arrived at the appointed time and place, I was a bit taken aback. I walked into an auditorium where the division was holding an annual professional staff meeting. When I walked in, the division director, probably out of courtesy, asked me to make a short comment to the several hundred IT professional staff attending the meeting in the auditorium. I was utterly unprepared and mumbled two minutes of nothing in my strong accent and broken English. (When I do not know what to say, my accent becomes stronger and my delivery of English becomes worse.) I was extremely embarrassed, to say the least.

This experience gave me a good lesson. From then on, whenever I go to a gathering, I always prepared a short elevator speech, just in case.

EXAMPLE: Using a three-point response to make up for an unprepared "elevator pitch"

Jon Huntsman, a popular former governor of Utah, ex-US ambassador to China, and ex-US ambassador to Singapore, is known for his three-point responses to almost any question. In responding to a reporter's query about this, he gave a three-point response: "It is easy to answer, easy to stay on top of, and easy to do" (p. 8, *Newsweek*, June 27, 2012).

Using the lexicon of this book, Huntsman's three-point response is essentially an unprepared "elevator pitch" to a spontaneous question. Here is my take on why the three-point response is an excellent way to deliver a good "elevator pitch" to an ad hoc question, to which we do not have a ready answer. (1) It buys us time to think of an appropriate response. While we are saying, "This issue has three points, the first point is . . . ," we have about 10 seconds to think of how to begin answering the question. (2) While we are expressing the first point, we gain additional time to formulate in our mind the second and third points. (3) Consequently, we have a better chance of leaving the impression that we are organized, knowledgeable, and intelligent.

Personally, I have used this approach to my advantage multiple times.

RULE 2:   Mastering a presentation by mastering the onset

Summarize the key points at the onset of a presentation or a written report.

Depending on presentation length, key points at the onset could mean 1 or 2 minutes to 5 or 6 minutes for an oral presentation, or one or two sentences to about one page for a written report. This helps the audience grasp the content and whets their interest in paying full attention to everything we have to say.

Furthermore, most people have a very short attention span, and a limited amount of time to give, whether it be listening to a presentation or reading a written document. It therefore behooves us to develop the practice of getting key points across with a minimal number of spoken or written words. If the presentation or document is meant to be short, these words are the only ones we need to express. Otherwise, these minimal words should serve as the summary, leading to a more detailed explanation.

EXAMPLE: Conference speeches

Have you ever noticed that when a speaker begins to talk at a conference session, everyone in the audience seems to pay close attention, but after a few minutes, although the audience is still looking at the speaker, their attention begins to wane? If you're listening to a good speaker, you'll see that he or she mitigates this drift by trying to make key points in those first few minutes.

Dr. Z is a highly regarded scholar and a member of the National Academy of Engineering. I've attended three seminars that he also happened to attend. During each one, he behaved the same way—he paid close attention to the speaker during the first few minutes, closed his eyes, and then opened them again during the Q&A period at the end to ask a pointed question. Seemingly, he always deduced his questions from what the speaker said during those first few minutes.

EXAMPLE: Newspaper articles

A common example of mastering the onset is something we see almost every day in a typical newspaper article. A well-written one begins with a headline and follows with a paragraph summarizing the whole story. After reading the headline, the reader decides whether to continue on to the article. If the headline is enticing, the reader continues to read the article. After reading the first paragraph, the reader then decides whether to read any further for more details. Sometimes, the headline or even the first paragraph is deliberately written in a way to entice readers to read further.

EXAMPLE: Emails

The writing style for a good news article can serve as a good model for emails.

In writing an email, the subject line is similar to the headline of a news article, and the first couple of lines are similar to the first paragraph.

The chancellor of a major university once told me that he receives tons of emails every day. He determines whether to read the full message in two steps: first, he judges the content in the subject line; then, if the subject line entices him to read further, he usually decides whether to read the entire email by what's contained in the first three lines.

RULE 3: Using three diagrams to simplify complexity

Use diagrams to explain complicated problems, but try to limit them to no more than three.

If a picture is worth a thousand words, then a diagram can be worth ten thousand. This is particularly true for complicated projects and systems. The trick isn't how to represent them with diagrams but how to represent them with only a very few of them—a few can add clarification, but too many would cause confusion. Regardless of a project's complexity, we can usually summarize it in no more than three diagrams. This task isn't easy, but it's important to provide a good conceptual overview as a tool to explain or market an idea to others, as well as to keep ourselves focused. In my own experience, as well as in my observations, this approach has resulted in big payoffs for those who can use it deftly.

EXAMPLE: Three diagrams to explain the subprime mortgage financial crisis

At the height of the subprime mortgage financial crisis in 2009, Frost & Sullivan, a financial and management consulting firm, gave a presentation to its clients about the global economic outlook, entitled "Global Economic Outlook: Bottoming Out Now, Recovery by June 2009."

The presentation used three diagrams to explain how the subprime mortgage problem led to the recession and financial crisis (www.frost.com/prod/servlet/cpo/154869931.pdf):

- One explained how subprime mortgage default problems led to the US economic slowdown.
- The second explained how this slowdown led to the international financial crisis.
- The third explained how the fear caused by the financial crisis led to recession.

I had read several long articles explaining how the subprime mortgage problem led to the financial crisis, but they still left me confused. These three diagrams gave a much better explanation than any of those long articles.

EXAMPLE: Three diagrams to impress a job interviewer

In 1993, when I was offered an interview for the political appointee position of Deputy Assistant Secretary for Information Systems at the US Treasury Department, I considered it to be a long shot. I knew I had to do something to distinguish myself, so I used the three-diagram paradigm.

A friend who was a vice president at a company doing a substantial amount of IT work with the Treasury Department lent me several documents describing its information systems. From these documents, I created three diagrams:

- One depicting how the communication system links the 14 bureaus that form part of the Treasury Department

- One depicting how the new technology may impact the communication system
- One depicting how the 14 bureaus may leverage each other.

Apparently, these three diagrams impressed the Assistant Secretary for Management, who interviewed me; they also gave him a better understanding of the Treasury Department's systems' current state and a vision of future growth. The Assistant Secretary offered me the job. While there were other factors that led him to make the decision, I believe the three diagrams helped a great deal.

EXAMPLE: Three diagrams to make a successful presentation to Congress

This is about how the IRS was able to use three diagrams to make the US Congress understand its multibillion-dollar revamp project of its IT systems.

In the mid-1990s, the IRS needed to undergo a multibillion-dollar upgrade of its IT systems. With such a large expenditure, Congress held regular hearings to monitor progress and to make sure the project was sound and the money wisely spent.

Such hearings typically began with testimony in the morning by the Government Accountability Office (GAO), followed by an IRS response in the afternoon. Although many of the GAO's criticisms were valid, some weren't. The main problem was that the IRS hadn't been able to articulate a convincing counterargument because it couldn't effectively explain what it was doing. The IRS would then get a beating down from Congress.

Shortly after a new Chief Information Officer took office at the Treasury Department (CIOT), with the responsibility of overseeing IT operations at the IRS, he began to look for why the IRS always got beatings at these hearings. Every time when he asked for a presentation on the IT revamp situation, the staff would arrive in groups of 5 to 10 and bring with them several volumes of documents. Of course, the CIOT could not comprehend what was going on with such large, voluminous, piles of information.

In desperation, the CIOT told the IRS staff he needed three diagrams:

- The first to depict the current system
- The second to depict the final system
- The third to depict the transition paths.

This was not an easy task, but after 10 months, the IRS was finally able to summarize its plan in three diagrams.

The timing was opportune: shortly thereafter, Congress called for a hearing. Just like in the past, the GAO made its critical comments in the morning. Before the IRS began its response, many observers expected the customary beating, but this time it didn't happen. The IRS was able to use the three diagrams to make Congress understand what it was doing and therefore deflected some of the GAO's criticism. For the first time, the IRS did not get a beating down at a Congressional hearing over its IT revamp project.

RULE 4: Sizing up and resonating with the audience

Assess the audience's background and interests, and fashion a presentation to resonate with the audience.

When we give a presentation, be it oral or written, we want to size up our prospective audience's background and figure out their interests. We can then use words, expressions, and lingo befitting our listeners, to make them feel we are conversing with them, not talking down to them. We can then mold the content of our presentation and hopefully induce a "that's right!" reaction from the audience.

For 30 years, "A Few Minutes with Andy Rooney" was a popular commentary made at the end of every *60 Minutes* episode on CBS. Rooney explained the popularity of his commentary in the following way: "I obviously have a knack for getting on paper what a lot of people have thought and didn't realize they thought. And they say, 'Hey, yeah!' And they like that." In other words, Andy Rooney knew how to size up his audience's interests and resonate with them.

It sounds obvious, but different audiences may resonate differently. We must tweak our presentation for different people.

Famed fashion columnist Diana Vreeland once said, "Pink is the navy blue in India." What she meant was that while blue is the most popular color in the United States, pink is the most popular color in India. The implication is that what may resonate with one group may not with another, and we have to adjust our perception accordingly in order to resonate with both.

Our elected political leaders are especially skillful at this—watch carefully what they say to different audiences, or in front of, say, their rural constituents versus their Washington, DC, colleagues.

Sometimes, we may have thought we sized up the audience correctly, but we could not get the resonance we hoped for. If we observe that their reaction to our presentation was a bland response, we may need to retune our presentation.

EXAMPLE: A tale of two gurus: one resonated with his audience, one did not

Two technical gurus worked at the same outfit. One failed. One succeeded. The failed guru tended to talk to his audience at his own technical level while the successful one made an effort to talk to the audience at the audience's level and to resonate with them.

Andy and Bob both have PhD degrees in computer science. Both are highly competent in IT. Both worked in the same IT group where their expertise stood out, and they frequently needed to make technical presentations to their colleagues. Before they joined this IT group, Andy used to work for a research organization where many staff held PhDs and had strong technical expertise, whereas Bob used to work for an engineering firm where few staff members held PhDs. Possibly due to this difference in background, they had different styles in making presentations.

When Andy presented a talk to his colleagues, he liked to use jargon and high-level technical terms. Bob was exactly the opposite: he tried to explain his presentations using words the audience could understand, thereby, making it easier to resonate with them. The colleagues really liked Bob, but were indifferent to Andy.

It goes without saying which one has been more successful in his career.

📖 EXAMPLE: A tale of two CIOs: one resonated with senior executives, one did not

A successful CIO should be able to use words his senior management can understand and slant discussions of issues to his senior management's interest.

Among the responsibilities that a newly appointed senior vice president (SVP) at a large organization assumed was supervising the CIO. The SVP didn't have any prior IT background and had difficulty following what the CIO and his operation had been doing. The only thing he knew was that the CIO and his group were responsible for the organization's computer and communication systems. He had no rapport with the CIO.

Several months later, he had an opportunity to hire a new CIO. Shortly after this, the SVP began to appreciate the mission importance of what the CIO's group was doing and became a strong advocate of the CIO's operation.

The difference between the old CIO and the new one was that the new CIO was more *sensitive* that his audience—namely, the SVP—wasn't technically oriented. So he spoke to the SVP in words that he could understand and used examples that matched the SVP's interests. The new CIO did have an advantage over the old one: he had been a consultant, and as such, he was accustomed to learning about his clients' interests and orienting his talk with expressions that his clients could understand.

📖 EXAMPLE: A tale of two cultures: one responded to a humorous analogy, one did not

The same comment drew totally different responses from two different cultures.

During some of Jason's technical talks, he needed to use certain results obtained from highly mathematical "queuing theory" analysis to explain a communication network phenomenon. These results are valid only under the condition called a "Poisson arrival process," a situation in which users arrive at a server (switches, transmission lines, bank tellers, and so on) "independent" of each other. To liven up his talks, Jason frequently explained that in spite of its imposing terminology, this pattern is quite common in real life, with the only exception being the lady's restroom. Ladies seemingly go to the lady's room in groups of two or three, so the "independence" assumption is no longer valid. This analogy usually breaks the ice and earns quite a few laughs, particularly from those who have never heard of the term "Poisson arrival process" before.

When Jason was on a technical exchange tour to China, he made the same joke, but it drew dead silence. Afterward, he asked someone in the audience why this was so. The answer: it was considered uncouth to use the term "lady's room" in an academic talk! The lesson, of course, is that people with different cultural backgrounds or working environment might respond (or resonate) differently to what we say, which we need to keep in mind when making presentations or engaging in conversations.

EXAMPLE: A tale of two readers resonating differently with the same book

Motivational books that combine well-known principles with good storytelling resonate extremely well in most readers' minds. A good example is the best-selling book by Malcolm Gladwell, *Outliers: The Story of Success* (Little, Brown, and Company, 2008). The book's core principle conveys a very well-known concept: your success depends on intelligence, effort, passion, and luck.

Interestingly, the book resonated with two of my friends quite differently. From what they said about it, you would never guess they were talking about the same book.

The first person, Winston, is a fairly successful senior engineering manager. What he got from the book is that your passion for what you do determines your level of success. The book portrays Bill Joy, a founder of Sun Microsystems, as totally immersed in and enthralled by what he does in the computer lab. Winston infers that people like Bill Joy can achieve what they have because they have an inherent passion, not just because they are smart or logged long hours at work.

The second person, Luke, is an average engineer; for him, the luck aspect (timing and environment) resonated more than passion, so he focused on the examples of people like Bill Gates and Steve Jobs; while talented and hardworking, they couldn't have been so successful without being in the right place at the right time.

The reason the two resonated differently is understandable. Winston is quite successful. Subconsciously, he wants to think his success is due to his intelligence and effort, not as much on luck. He readily admits that he does not have Bill Joy's passion, nor does he want to lead a life like Bill Joy's. But that is his personal choice, not luck. To him, intelligence and effort is far above luck. Luke, on the other hand, has not been as successful as Winston. Subconsciously, he wants to think his lack of success is not due to his lack of intelligence or effort, but his lack of luck. To him, luck is above everything.

RULE 5: Being careful of careless comments

*Be mindful of what we say. Carelessly worded comments can lead to
negative reactions.*

After a big win in Florida's primary in January 2012, Mitt Romney was apparently very happy and confident. Early the next day, he candidly spoke to a reporter, saying, "I'm not concerned about the very poor. We have a safety net there. If it needs repair, I'll fix it." In no time, his rivals took the statement, "I'm not concerned about the very poor" out of context and piled criticism on him. Clearly, he didn't deliberately throw red meat to his rivals or aim to alienate voters, but that's exactly what happened.

Most unguarded comments are harmless. But when one of them does stir up negative feelings, the backfire can be serious. Unguarded remarks can inadvertently rub people the wrong way; stir up a strong negative reaction, and may cause severe damage to our objectives.

Gratuitous comments often follow overconfidence. Being wary of overconfidence can reduce the occurrence of gratuitous comments.

I know of several very smart people who could have had more successful careers if not for their propensity for uttering insensitive, often offensive, comments. (An example is given below.) I am also aware of individuals who failed their job interviews due to careless gratuitous remarks. (Examples appear in Chapter 7, Job-Interview Smart.)

EXAMPLE: Insensitive comments blunted career advancement

Dan was a very smart and knowledgeable computer scientist. He had been working successfully as a senior technical staff at a think-tank when the director of an engineering division from another company recruited him. (In practice, a think-tank is an expert consulting firm.) Dan's new position was still as senior technical staff, reporting to a second-level manger. But the director promised Dan that he would be made a division fellow if he worked out after a certain period of time.

Dan worked well with on all his technical assignments. Unfortunately, he had one problem: he liked to make gratuitous comments about the difficulties of technical problems, saying words like, "It's so simple," "It's easy to understand," and so on. When Dan was working at the think-tank, he might have to show off to his clients that he was knowledgeable and was capable of solving any problem. But when he acted this way with colleague in the engineering division, he appeared to be arrogant and belittling. As a result, he was not very popular. Dan's manager described the situation to the division director. Consequently, the director decided not to make Dan the division fellow.

I cannot believe he deliberately said things to upset others. He was just being careless in choosing his words.

📓 **EXAMPLE: A careless remark resulted in a lasting negative impact**

By any account, General Alexander Haig was a high achiever. Unfortunately, he's most remembered not for his achievements but for an unguarded statement.

In the military, Haig was made a four-star general in his 40s and rose to become the supreme commander of NATO. In the civil service, he was chief of staff at the White House, and eventually secretary of state. In industry, he was CEO of United Technology.

On March 31, 1981, President Reagan was rushed to the hospital from a bullet wound inflicted in a failed assassination attempt. Shortly after, Haig, then the secretary of state, called a press conference at the White House. He made the following unguarded statement: "As of now, I am in control here in the White House." His intention might very well have been to communicate that all was well. But, whatever his real intention, many people viewed this statement quite negatively. On the surface, at least, it implied he had taken over control from the president when he should have stated that the proper chain of command was still in place, or that President Reagan himself was still in control and all was well.

He never lived it down. As a political figure, he was a damaged good. A little more than a year later, he resigned and never took up another government position. He tried to run for president but couldn't get his party's nomination.

One has to wonder whether his statement had any anything to do with his failure to win the nomination. Could history have been different had he not made that unguarded statement?

RULE 6:   Using plain language

Use words and expressions that our intended audience can readily understand.

Anyone who has seen the movie Margin Call is likely to have come away remembering the expression, "Explain it to me in English." The movie is a story about a fictionalized Lehman Brothers-type stock brokerage firm. A financial analyst, using a mathematical model, suddenly realizes that the company is in imminent danger of collapse. He has to explain this to his first-level manager, to his second-level manger, and all the way up to the CEO. At the beginning of each of these meetings, he is told right away to "say it to me in English." The CEO even asks the analyst to explain it as though he were talking to a child.

A large corporation hired a new COO (chief operations officer) from outside the company. Upon his arrival, he called a meeting for all the divisions to brief him on their activities. The first presenter began this presentation with an acronym that was well known within the company but not to the new COO. The COO asked the presenter to expand the acronym, which he did. As the presenter resumed his presentation, subconsciously he threw in a couple of additional acronyms. At this point, the COO abruptly stopped the presenter and said, "That is enough for today." He then asked the next presenter to continue the briefing. From then on, every presenter was careful not to use acronyms, but for the first presenter, the damage was already done.

The lesson is clear. Only use words and expressions that are commensurate to our audience's background. Avoid using jargon, acronyms, and technical terms unless we can be sure that our listeners understand them.

EXAMPLE: Rising to a senior position at a young age

In 2013, Brian Deese was appointed as Deputy Director of the Office of Management and Budget (OMB) at the age of 34. (OMB is arguably the most powerful agency in the federal government. It controls the budget and oversees the management of all federal agencies.)

It is quite impressive, if not phenomenal, for Deese to rise to such a powerful position at such a young age. He must have been very smart and capable, but there are others as smart and capable among President Obama's assistants. What made Deese such a standout? I believe the answer lies in the following statement, which appeared on Reuters.com on March 4, 2013: "Deese is known within the White House for being able to explain complicated economic policy in layman's terms. He has traveled with Obama on domestic trips as the president pressed for economic policy initiatives." That is, he is skilled in using plain language to explain economic policy to Obama and to stakeholders.

EXAMPLE: Communicating with management

It is our responsibility to make the boss knowledgeable and to understand what we are doing.

Judy, a first-level manager at an electronics firm, once sought advice from me. She complained that her direct manager didn't understand or appreciate what her group was doing. I asked whether her manager held regular staff meetings, and she responded that he did. I then asked her whether she made presentations during those meetings about her group's activities. When she again replied in the affirmative, I advised her that this miscommunication was actually her fault. She violated two rules of "smart soft skills": "Focusing on self-examination, rather than putting blame on others, when things gone awry" (see Chapter 4, Work Smart), and "Using plain language." It was her responsibility, not her manager's, to make her manager understand what she and her group were doing. She just had to improve her communication skills. I suggested that she should, using plain language, practice explaining to her mother, who had no technical background, what her group was doing. If she could make her mother understand after a 5- to 10-minute presentation, she should be able to make her manager understand as well. A few months later, she called and thanked me for the advice and told me that I was right.

🔲 **EXAMPLE:** Turning plain English to an advantage

I have always used plain English to explain any problem. To me, I have no choice. Coming to the United States as a graduate student at the age of 24, I could hardly speak English. The only way I know how to explain anything is to use simple, plain words. I wish I could speak with flair, but I can't. By luck, this handicap of mine has turned out to be a blessing.

During the period when I was actively engaged in doing consulting work, a client named Sunil introduced me to a new client. Sunil advised me to have my first technical discussion with this potential new client in person, not on the phone. He told me frankly that, on the phone, with my accent, people could not understand me well. But in person, I have been able to use plain English to make people understand complicated technical problems. Apparently, I made some of my clients happy because I could explain technical problems to them using plain words.

During the farewell party at the end of my tenure at the US Department of Treasury, my boss paid me a special compliment, saying that he had always understood my presentations at the weekly staff meeting (he is nontechnical). The implication seemed to be that his other direct reports could not do as well. When one of my direct reports made a similar comment, my boss led a standing ovation. Later, when my boss moved to head another agency, he wanted me to work for him again. (I had to decline, for personal reasons.) Apparently, my way of using simple English to explain things actually helped my career.

RULE 7:   Using jokes and self-deprecating humor

Have a sense of humor and humility by poking fun at ourselves and mixing jokes with our presentations. It helps to draw audience attention.

We all know that an interesting presentation needs to be inflected with humor and jokes. This is particularly so if we observe a bland response from our audience. The easiest audience-pleasing jokes are those that poke fun at people by stereotyping them. However, if we make such jokes, listeners may laugh, but they will not necessarily respect us for making fun at someone else's expense. By doing so, we end up lowering our own esteem and offending people we have stereotyped. What we should do instead is make fun of ourselves. By doing so, we not only add humor to our speeches without offending others but also exhibit our own humility. We all have enough weak spots and/or peculiarities to joke about. Take me, for example: an ethnic Chinese, a researcher, a professor, a consultant, a manager, and one who speaks with a strong accent. There is more than enough material for me to poke fun at myself.

In addition to self-deprecating humor, we need to inject other jokes into our speeches as well.

One important point we need to realize is that these jokes by themselves do not add much value, if at all, to our presentations. They become an effective tool to make our presentation more interesting and help draw the audience's attention to what we have to say, only if they can be associated with the subject and contents of presentation.

The benefit of self-deprecation is not limited to making presentations. A modest amount of self-deprecation, when used sparsely in regular conversation, can make us more likable and project an image of being successful and confident. (Being people smart!)

■ EXAMPLE: **Backfire caused by stereotyping others**

A large high-tech company was establishing a new research center for communications. To celebrate, the center organized a workshop and invited many well-known researchers in the field. For the dinner banquet, many senior executives and local dignitaries were invited as well. The "communication" here involves electronics and electromagnetic waves, so the center director thought it would be interesting to have a dinner speaker who is a speech communication expert to talk about "communications" from a totally different perspective. (Note: this is the type of communications this chapter is addressing.)

The speaker, Cassidy, was to give a talk on the subject of miscommunication. As a seasoned speaker, he started with a joke.

The joke belittled two ethnic groups. In particular, he stereotyped ethnic Chinese from old Western movies/TV shows as menial workers, uneducated, and ignorant. While he was successful in bringing out laughter from the audience, one of the two ethnic Chinese in the audience stood up, expressed his displeasure of Cassidy's joke, and walked out. The remaining talk ended in a quite unpleasant mood.

(Lessons learned: (1) Do not make ethnic jokes in public speeches. (2) If we do make them, do not make insulting ones.)

■ EXAMPLE: **Using self-deprecation to liven up presentations**

Injecting into our speeches self-deprecating jokes that can be linked to the subject of the speech is a particularly effective way of making our talks more interesting and enjoyable to the audience.

I stumbled on this self-deprecating technique out of necessity. I face two handicaps when I make a presentation: a typically boring technical topic and a heavy accent. This combination is a sure recipe for boring everybody in the audience, a harsh reality I discovered during my first presentation at a conference. At that time, I knew I had to say something humorous to get the audience's attention. But I was not able to think of any appropriate joke on the spot, except for poking fun at myself. The topic I was presenting concerned communication protocols, so I pointed out that a communication protocol is a set of language rules that have been defined by its creators. Different protocols are created by different people and therefore have different rules. I then related an embarrassing story in which I mixed up a certain protocol in Chinese language with that of English. This self-deprecating joke brought laughs from the audience and livened up my talk. (Author's note: I removed the specific joke from the book under the suggestion of a reviewer. He did not think I should put an ethnic joke in the book.)

EXAMPLE: Taking cues from our political leaders

In his victory speech on Election Day in 2008, US president-elect Barack Obama referred to himself as a "mutt." In her swearing-in speech at the Department of State, Secretary Hilary Clinton thanked her husband for teaching her all kinds of "experiences." During his guest appearance in February 2013 on the *Late Show with David Letterman*, New Jersey Governor Chris Christie poked fun at his weight and called himself the healthiest "fat guy." If our leaders are humble enough to make fun of themselves, we common folks should be able to do likewise.

EXAMPLE: A survey on sense of humor

Accountemps, a large employment agency, conducted a survey of more than 1400 CFOs (chief financial officers) about the question, "How important is an employee's sense of humor in him or her fitting into your company's corporate culture?" Approximately 79% of them said it is important (accountemps.rhi.mediaroom.com/funny-business). In reporting the story on its website on June 2, 2012, MSNBC asked visitors similar questions, and almost 97% of them gave a positive response.

Surely, a person capable of self-deprecating is likely to have a sense of humor. So we should be sure to add self-deprecating humor and humility, as well as other humor, to our speech or conversation. After all, it is simply being people smart!

PART TWO

Dealing with People
THE ESSENTIAL

CHAPTER TWO

People Smart

Principle: Making people feel good

Strategy: Put ourselves in other people's shoes

For many people, US President Bill Clinton's best-known quote is, "I feel your pain." By this, he meant he knew how people felt as though he were in their shoes. His empathy and compassion for people's problems translated into loyalty from many voters. President Clinton is the quintessential example of being people smart.

If we want people to accept us as we are, we must first accept them as they are. If we want to deal with people successfully, we must make them think that they are successfully dealing with us. If we want people to like us and be willing to work with us and help us, we must behave in a way that would motivate them to do so.

These arguments share a common thread: knowing human nature well can enhance our chances in developing winning strategies that are people smart.

Practicing the following rules, defined later in this chapter, can help us achieve deftness in being People Smart:

- Getting accepted by accepting others first
- Winning by understanding both ourselves and our counterparts
- Being aggressive by being nonaggressive
- Gaining by giving
- Successful networking by networking less

Fast-Tracking Your Career: Soft Skills for Engineering and IT Professionals, First Edition.
Wushow "Bill" Chou.

▦ Being heard by listening

▦ Getting liked by using self-deprecating humor in conversations (refer to Chapter 1, Communications Smart, for a general discussion of self-deprecating)

▦ Avoiding blunders of overconfidence (refer to Chapter 4, Work Smart, and Chapter 7, Job-Interview Smart, for a general discussion of "overconfidence").

RULE 1:   Getting accepted by accepting others first

Accept people around us by assimilating into their culture. This is the easiest way to build rapport with people around us and to be accepted.

When Secretary of State Hilary Clinton arrived in Bangladesh during her whirlwind diplomatic visits to several countries during May 2012, the media made her visit front-page news. The story was not as much about any diplomatic achievement but about Clinton's makeup, or rather, the lack of it. Commentators had a field day second-guessing the reason for her wearing little makeup, but none of them seemed to have pinpointed the real reason. As I see it, Clinton was just trying to assimilate into the local culture. In Bangladesh, respectable women wear little makeup. In particular, the key people Clinton was to visit were the country's prime minister, Sheikh Hasina, and foreign minister, Dipu Moni, both of whom are women who wear little makeup. When the three of them posed together for a photo, they looked comparable. Had Clinton worn her usual makeup and coiffed hair, she would have looked out of place. (For a picture of the three posing together, see www.worldmeets.us/bdnews24000001. shtml#ixzz1yBqhw4FM.)

We want people around us to feel that we are one of them. But we must accept them before we can expect them to accept us. The best way to achieve this is by assimilating into their culture.

Every working environment has its own culture. To fit into that environment, we need to assimilate into it: "In Rome, do as Romans do." We need to think and behave as they do, in terms of, but not limited to, their use of words and lingo, dress codes, protocols, mannerism, humor, social activities, and ideologies.

▣ EXAMPLE: Words and lingo

Each environment has its own vocabulary of words and lingo that it considers proper or improper.

W. Scott Gould, Deputy Secretary of the Department of Veterans Affairs under President Obama, always speaks properly and eloquently. One early Monday morning, I attended a meeting he chaired when he was Deputy Assistant Secretary for Management at the Department of Treasury. To my utter surprise, he mixed his remarks with four-letter words. Prior to the meeting, I had never heard such language from him. Later that day, I learned that he had been on Navy Reserve duty the previous weekend. Apparently, when he worked at his regular job, he spoke in a way befitting an upwardly mobile senior executive, but when he was on Navy Reserve duty, he spoke like a sailor. In our Monday morning meeting, he hadn't quite converted back to his normal speaking pattern.

EXAMPLE: Dress codes

Each environment has its own style of dressing that it considers proper or improper. How we dress can help us build rapport with people around us or make us look out of place.

Graham was a professor with both teaching and administration responsibilities at a university. He also did some consulting for industry. When he went to classes or otherwise interfaced with students, he wore jeans and other casual wear, just like how students dressed. When he went to university administrative meetings, he wore a baggy suit, just like typical university administrators did. When he visited his clients as a consultant, he wore a designer suit to project an image of authority and success. Once he had to visit Singapore in a professional capacity. When he arrived at his client's office, nobody initially realized that they had an American visitor because such visitors usually wore jackets and ties, but Graham didn't—he wore a short-sleeved white shirt without a jacket and tie, just like the local professionals dressed. This helped him build a good rapport with his clients immediately.

However, some people seem to enjoy being contrarians, possibly as a reverse status symbol. I know an extremely smart person who takes pleasure in dressing sloppily. He's a member of the National Academy of Engineering and a wealthy inventor, but he always wore the same wrinkled khaki jacket. He once proudly told me that he owns only one jacket—the one he was wearing—but on a tour of his mansion, his wife showed me around their living quarters, where I spotted a closet that happened to be open. In it, I saw five or six of these identical khaki jackets! Apparently, for successful techies, this is a status symbol. Facebook Chief Executive Officer (CEO) Mark Zuckerberg wears the same grey T-shirt every day, and the late Apple cofounder Steve Jobs wore a black turtleneck all the time, just to name two.

EXAMPLE: Military versus civilian culture

People with a military background tend to be more formal with respect to protocols and reporting procedures between staff members with different ranks, whereas people with nonmilitary backgrounds, especially high-tech ones, tend to be less formal. If we need to work with both types of people, it's clearly advantageous that we be adaptable to both.

Lucy, a civilian administrative assistant working in the federal government, was temporarily assigned to a group that consisted of both military and civilian staff. One day, she noticed a typographical error in a report that was to be forwarded to her civilian boss. Instead of forwarding the report as it was, she casually mentioned the typo directly to its author—a colonel—to speed up the correction process. Instead of showing appreciation, the colonel dressed her down for not following the proper protocol. He didn't feel a lower-ranking staff from another chain of command should have talked to him directly. In particular, she didn't show as much deference to him as he was used to getting from lower-ranked officers.

🖥 EXAMPLE: The art of a smile

Being able to beam a congenial smile at any moment is a useful trait in any organization. (It sounds easy, but it is not. I have not yet learned how to do it.) It is particularly beneficial if you are working for the government.

Within governmental organization structures and promotion processes, it is crucial that no one badmouths you if you want to advance to a senior managerial position. Therefore, it almost becomes a culture that most senior career managers are quite congenial and ready with a nice smile at any moment. (This is my personal observation; others may or may not agree.)

Once, I attended a social party at which I noticed a gentleman, Jesse, smiling pleasantly and working his way through the room. To me, his mannerisms typified a senior career government official, which I disclosed to a friend sitting next to me. My friend knew Jesse and confirmed that he was indeed a high-ranking government official.

RULE 2:   Winning by understanding both ourselves and our counterparts

Understand the strengths, weaknesses, likes, and dislikes of ourselves as well as those with whom we are dealing. This provides us with the information we need to develop successful winning strategies.

The statement "understanding both ourselves and our counterparts" is a metaphorical interpretation of a quote from the book, The Art of War, written by Sun-Tzu, one of the best-known war strategists. The literal translation of the quote reads, "know thyself, know thy enemy, and you shall win every war." While this statement was intended for fighting wars, we can apply it metaphorically to our professional lives as we try to build a successful career: "know yourself, know your counterpart, and you shall have a winning hand."

🖥 EXAMPLE: Website considerations

In today's workplace, an effective Web presence is mission-critical. To launch a successful website, we must know well both our (1) organization's objectives and (2) targeted clients' interests. The goals are to entice our targets to visit our website instead of our competitors'; to promptly and intuitively provide the information our targets seek; and to offer our targets a rewarding experience in accessing our website.

But, alas, some websites are still poorly designed and not able to effectively meet the organization's objective and satisfy clients' interests.

Here's a commercial website example based on my own experience as a user. Two upscale competitive hotels in Los Vegas, A and B, were offering deep discounts during periods of low occupancy. I clicked on Hotel B's advertisement, and a website popped up immediately with a calendar associating different discounts for specific dates. I could easily and quickly determine which dates to choose and make the appropriate reservation for the specific discounts that interested me. As a comparison, I clicked on Hotel A's advertisement. A webpage with a meter popped up, to show the amount of loading time I could expect to wait. After a couple of seconds, the homepage appeared, but it was wider than the window screen, so I had to use the mouse to shift the page left or right. When I finally found the reservation button and clicked it, a third page popped up, asking for my arrival and departure dates. Surprisingly, when I input the dates, a fourth page appeared, displaying a room charge that was substantially higher than the advertised one. Apparently, the dates I wanted didn't have discounts available, but regardless, I had no way of knowing which dates *were* available for discounts, if at all. Obviously, my choice was Hotel B. From my perspective, the people who designed Hotel A's website didn't know the hotel's objective (selling rooms at discounts during days with low occupancy) or the potential customer's desire (finding those discounted rooms easily). They designed a site that didn't entice anyone. They lost one potential customer, likely many more.

I travel a lot and always make reservation on flights, hotels, and even tour guides over the Web. On some of these websites, the entire process is a cinch. But on others, it is unbelievably tedious and confusing.

Everyone knows that Google's success is due in part to the simplicity in its Web design. Yet there still are Web designers/CIOs who seem to have paid no heed. They either do not understand their corporate objectives and their customers' interests, or they are simply incompetent. (The person who was responsible for Google's simplicity is Marissa Mayer, a computer engineer by training. Yahoo recruited her from Google. She became Yahoo's CEO at the age of 37.)

EXAMPLE: Turf tussle considerations

Most middle and senior managers seem to have a strong common desire in acquiring more responsibility and power than they already have. If necessary, some would even stab their friends in the back. If we face such possibilities, we need to assess our own situation and be prepared for such a possible onslaught. That is, know ourselves, know our colleagues, and be prepared.

I recently talked to a senior VP of a large corporation and asked him point-blankly whether he had experienced any backstabbing. His answer: "Of course! And I have stabbed others in the back as well. Sometime, the stabbing even comes from the front. You should expect it. Such actions come with the territory." He ought to know; he had just been badly backstabbed and lost much of his turf. His career had been on a phenomenally fast track. In about 7 years, he had been promoted from a first-level manager to second-level manager, director, senior director, VP, and finally a SVP in charge of five business units. I assumed he probably had stepped on many toes and stabbed a couple of backs. Apparently, he had been busy in expanding his turf and getting himself promoted and his power expanded, but he had not paid attention to the art of preventing being stabbed. He made a preventable misstep (he hired a personal friend to a senior position) and gave his opponent the opportunity and ammunition to "stab" him. He understood himself, he understood his adversaries, but unfortunately for him, he did not use this information to prevent a misstep.

(*Lesson learned: We should be careful not to do anything that could be used by our adversary as a "knife" to stab us in our back.*)

📓 EXAMPLE: Outwitting a strong hand

One evening during a professional conference, a group of conference participants goes to a sports bar for drinks. Among these were Eddy and Barney. They knew each other well. Eddy used to be a football player: he is macho and can hold his liquor. Barney is the opposite: he is not macho and has a low tolerance for alcohol. After a couple of drinks at the bar, Eddy challenged Barney to down the glass of wine each was holding in one gulp. Barney, knowing Eddy's macho personality, responded by saying, "We all know you are a good drinker. Yet knowing that I cannot drink much, why do you want to humiliate me? If you want to show us you are a real good drinker, a fair-minded person, and do not intend to humiliate me, you should offer to fill your glass up with vodka instead of wine. If this is your offer, I am willing to accept your challenge." Eddy fell into Barney's trap and agreed. Drinking down a glass of wine made Barney's face flush. But drinking down a glass of hard liquor made Eddy really drunk, and he threw up. Barney knew his strong point (being clever), his friend's weak point (being macho), and a winning strategy. To boot, Barney won admiration from others at the same gathering for his quick wit.

(*A caveat:* Sometimes in winning a battle, one could lose a war. Barney almost lost a friend.)

RULE 3:   Being aggressive by being nonaggressive

Maintain a nonaggressive attitude even if we mean to be aggressive in pursuing our career advancement. We can achieve more with less aggressive behavior.

The right way to achieve our career goals aggressively is by desisting from aggressive behavior. Without appearing to be aggressive, our colleagues and bosses are not guarded against us, are willing to work closely with us, and will give us the benefit of the doubt.

Many motivational speeches, articles, and books urge us to be assertive and aggressive. Such a strategy could be correct for people who are content to stay at a low- to middle-level position, but it is definitely wrong for people who desire senior positions. Most people are turned off by those who appear to be aggressive. The reality is that by being outwardly aggressive, we might make a gain in the short term but lose in the long term. (Winning a battle, but losing a war!)

I always advise young people not to demand a pay raise or a promotion. It's in the manager's best interest to distribute his resources in such a way as to ensure that good workers are happy and rewarded properly, so arguably the best way to get the recognition we seek from a boss is to improve our job performance to his/her liking. (If, unfortunately, we work for an unappreciative or an inept boss, the best strategy is "to stop polishing the sneaker" and find a way out. See Chapter 6, Career Smart.)

Some individuals do get a pay raise or a promotion by persistently pleading with their managers. But doing this more than once or twice will leave a bad taste with their bosses. It could take even longer to get the next pay raise or promotion. I know of someone who finally got a promotion in his grade after repeated pleading with his manager. But that was over 10 years ago, and he has been at the same grade ever since. This is not the way to be on the fast track!

EXAMPLE: The last pay raise

Justin was cleverer than many in his approach to get a pay raise. He did not plead with his boss. Instead, he applied for a position at out-of-town companies, and when he got an offer that was higher than his then salary, he approached his boss. His boss did not want to go through the trouble of finding a replacement for Justin's position, so he agreed to match Justin's salary with that offered by the other company. Justin was very pleased with his aggressiveness. He had a free paid trip (for the job interview at the other company) and a good pay increase.

Because Justin's salary, after the raise, was higher than colleagues of equal skills, he did not get a pay raise for three years (except for a cost of living adjustment). Justin felt it was time for him to repeat the same strategy again. When he again got an offer with a higher salary, he went to see his boss, but this time, to Justin's surprise and disappointment, his boss said, "Best wishes!"

EXAMPLE: A tale of two new hires: one outwardly aggressive, one not

An openly aggressive new hire may get a higher starting salary but could end up with less in the long run.

A manager needed two additional staff members. Among the candidates screened through the Human Resources department, only two met his needs, Matt and Nancy. The manager made offers to both. Nancy accepted the offer right away, but Matt demanded a higher salary than the one offered. Because the manager needed two staff right away, he didn't have time to go through another hiring process to search for another qualified person, so he reluctantly agreed but wasn't happy about it. Both Matt and Nancy were good workers and did equally well, but because Matt had a higher salary, the manager compared his performance with those of senior staff and evaluated accordingly. Matt wasn't on par with the senior staff members and thus didn't get a high rating for his performance. The manager evaluated Nancy as a junior staff member, and, among this group, she did extremely well, so she received a higher performance rating, a larger salary increase, and a bigger bonus. Nancy's base salary was still lower than that of Matt, but with the bonus, Nancy's overall compensation was actually higher. In addition, Nancy's personnel file indicated a better performance rating than Matt's. Matt might have won initially for being aggressive, but he ended up getting less, and the lower performance rating may impede his promotion in the future.

📖 EXAMPLE: Nonaggressiveness facilitates a career

I know of a person who never asked for a pay raise, never asked for a promotion, never negotiated for a higher starting salary, never requested more authority, and always shared or gave credit to his staff or his boss. He may have appeared to be not aggressive. But all his strategies, whether inadvertent or intentional, had the same effect as being aggressive without any of the attendant disadvantages. Throughout his career, he was able to receive higher salaries, get faster promotions, gain more responsibilities, and earn greater loyalty from his staff than his peers with stronger backgrounds.

RULE 4:   Gaining by giving

Harvest professional favor from our friends by doling out favor upon them first.

Doing a favor to people does not cost us much, in terms of effort, time, and/or money. But the value of potential returns, in terms of professional connections, could be invaluable.

I use the term "favor" here in a very broad sense. It includes material generosity, such as sending gifts and paying dinner bills, for appropriate occasions. It also includes nonmaterial generosity, such as

- Being respectful with respectful temperament
- Being magnanimous
- Being kind
- Being helpful
- Sharing credit (this generous act will be discussed further in Chapters 8 and 9 on Boss Smart and Motivating Smart)

- Being considerate
- Showing appreciation
- Giving compliments
- Mentoring
- And so on.

Nonmaterial generosity is usually far more valuable than material generosity.

Our generous act should be genuine, not calculated to expect specific returns. What we gain is a good reputation and an "I owe you" from people who have received favors from us. Often, in unexpected opportune occasions, people who either have benefited from our generous act or are aware of our reputation could be in a position to give a strong reference for a position, reveal an inside track for closing a deal, or be otherwise helpful in a variety of possible situations.

On the other hand, a calculated generous act that expects certain returns, once detected, may not be appreciated by the recipient. It might even backfire.

Admittedly, most generous acts do not result in any return. But for those that do, the value could be immeasurable.

EXAMPLE: Being respectful

Being generous with our respect to others by doling out proper respect to people we are interfacing/conversing with is probably the most beneficial tactic of being generous. This can help us maintain a long-lasting good rapport with our bosses, our staff, and our counter parts.

However, to maintain an appropriate respectful temperament and mannerism to all people and on all occasions is not an easy task. I know I cannot do it. But I do know one person, Jin-Fu Chang, who excels at this skill and benefits greatly from it.

Dr. Jin-Fu Chang, an IEEE fellow, is an electrical engineer by training. As of this writing, he is the president of a university in Taiwan. In addition, he has been the minister for science and technology and the president of another university. I asked him what soft skill stands out as the most important in his careers. He said, and I am paraphrasing, "being respectful with respectful temperament."

When he interfaces with his superiors, he listens attentively, but not obsequiously. He would pay particular attention as not to waste their time; and would make his presentation concise.

When he interacts with his staff, he never talks down to them and he is always attentive to what his staff presents to him.

When he attends any meeting he always exhibits a calm manner even if he has just come out of a depressing and stressful meeting.

(Of course, with Dr. Chang's highly successful career, he is good at many other soft skills as well.)

▉ EXAMPLE: Being magnanimous

President Obama is an excellent role model for being magnanimous. He has always been willing to work with his adversaries. Usually, it costs him nothing, but the return, when there is one, is immeasurable. Some might disagree with him on ideological issues, but putting that aside, he's been able to get extremely capable people who were his adversaries and rivals to serve him as vice president (Joe Biden), secretary of state (Hillary Clinton), and other important positions.

▉ EXAMPLE: Being kind

A large corporation experienced a management turnover. One senior manager, Ed, lost all his managerial responsibilities and staff. He was not let go, but he was made a staff member and assigned as an advisor to a senior vice president (SVP). Because everyone on the new corporate management team was busy, nobody was particularly friendly to him, except for one vice president, Frank, who treated him with the same respect and friendship as someone who was still in power, not someone who had lost it. It just so happened that the SVP assigned Ed to represent him on the corporate management committee. Ed sometimes passed on to Frank what happened in these committee meetings, giving him an advantage.

Ed also helped Frank on another matter. Any senior manager knows the important role a good administrative assistant can play. Ed knew the best administrative assistants in the company and which of them wanted to work for a different boss. So, with Ed's help, Frank recruited an excellent administrative assistant from within the company. Frank's primary investment in all of this was just being nice and friendly, even when someone was "down," and the returns in this instance were inside information and recruitment insight. Both were invaluable and helped him tremendously.

▉ EXAMPLE: Being materially generous

I've known quite a few people who host parties for friends and acquaintances, try to take care of the bill when dining together at restaurants, or send cards and gifts on special occasions. By doing so, they've established a good rapport with their friends and acquaintances and are well liked. I do not have solid evidence that indicates these generous acts have helped their careers, but it's difficult to imagine that they haven't. If such people were generous to you, and you happened to know some information that they would find useful, wouldn't you want to pass it along to them as long as doing so didn't violate ethical or legal rules? Or if someone asked you about them, wouldn't you more likely emphasize their strengths and downplay their weaknesses?

📱 EXAMPLE: Being helpful

Most people in the IT field probably know of Leonard Kleinrock, a member of National Academy of Engineering and a recipient of many other honors, known either as one of the two coinventors, or as one of key inventors, of the Internet. During the late 1970s, he regularly hosted a short course, with him, Larry Roberts (also a member of National Academy of Engineering and known either as one of the two coinventors, or as one of key inventors of the Internet), and a third equally illustrious individual as instructors. Shortly before one of the scheduled tutorials, the third instructor couldn't make it, so Kleinrock asked me to fill in. I agreed and did a conscientious job. I just wanted to be helpful, hoping not to embarrass him or the tutorial. I was not expecting anything in return. (Yes, I was paid. But from my perspective, this was not consequential.) To me, conducting the same tutorial with two Internet inventors was rewarding enough, but Kleinrock apparently felt he owed me one. He has since written reference letters for me, recommended me for consulting work, and done many other things that have increased my professional prestige and visibility.

I have done many helping acts throughout my life. Most of them did not get anything in return. But when there is a return, its value could be immeasurable, as exemplified by the above story.

📱 EXAMPLE: Being grateful

A thank-you note might not seem like very much, but it could make a big difference. Randy Pausch told a story in his bestselling book, *The Last Lecture*, about one such situation. He was on the faculty at Carnegie Mellon University. One semester, he and another faculty member were reviewing student applications. In one student's file, they found a thank-you note from the student to the secretary, thanking her for her help in the application process. Duly impressed, they accepted the student. (Randy Pausch was a computer engineer by training. For background about him, see http://en.wikipedia.org/wiki/Randy_Pausch.)

RULE 5: Successful networking by networking less

Establish a successful network by building upon our own visibility and credibility in our professional community.

The degree of benefit we get from networking depends not on how many people we know, but on how we are known in terms of our professional and personal reputation and on whom we are known to. Successful networking means knowing the right people who can and will speak out or do things for us at the right time and in the right place. They can, because they are the right people in the right positions; they will, because they're aware of and respect our skill and ability, owe us a favor, and/or have a good rapport with us.

Many people mistakenly think that networking simply means knowing a lot of people. From the perspective of career growth, the effort spent on networking with superficial acquaintances has practically no value and is a waste of time.

We can have two types of network connections: real and virtual. The "real" network consists of those we know personally. The "virtual" network of professional contacts consists of those whom we may not know personally or know only superficially. But they know us by our reputation or are familiar with our activities.

I have not heard much talk about virtual networking. But for people at senior levels, virtual networking plays a very crucial role, often more so than real networking. This is particularly true if one applies for a senior-level job. It is common practice for a recruiting committee to solicit input from people in the candidate's professional field, beyond the list of references provided by the job applicant. In addition, virtual networking can also play a very crucial role for other professional activities.

EXAMPLE: Participating in cross-department studies and committees

Taking a leadership role in cross-department studies and committees can extend our visibility to upper management and other departments. This has the effect of expanding our real and virtual networks throughout the company.

A certain very large corporation consists of multiple divisions, each with its own independent IT shop. The company also has an interdivision IT workshop. One year, the workshop was looking for volunteers to carry out two studies. One was on defining metrics and procedures that could be used to measure the performance and progress of IT projects. The other was on corporate-wide standards that could be used for purchasing hardware and software.

Janice, a senior IT manger, volunteered to lead the study on the IT metrics, and Sam, a senior technical staff, volunteered to lead the study on standards. They both did a wonderful job, and their works were well received. By taking on these tasks, they both broadened and enhanced their visibility and substantially expanded their virtual network within the corporation. Within a year, Janice was recruited by another division as its CIO, and Sam was recruited by a third division as a senior manager.

EXAMPLE: Giving speeches

Many people believe it's hard to get a speaking engagement. Actually, it's quite easy as long as we have some useful professional information to share. Conferences and professional meetings are always looking for knowledgeable speakers.

Giving speeches at conferences and professional meetings is the easiest and best way to establish a successful real network of contacts, as well as a virtual one.

We all have a tendency to perceive a conference speaker as an expert, especially if we've seen the same person's name appear at several different meetings. We've also unknowingly become part of this speaker's virtual network: if we ever need to identify someone with the expertise associated with that person's speech topics, guess who we'll remember?

I'm personally aware of several frequent conference speakers who have been approached by strangers and offered excellent positions or consulting contracts. I'm also aware that strangers have provided strong recommendations to some of these speakers.

EXAMPLE: Publishing papers

Publishing articles frequently has a similar effect to speaking at meetings and conferences and is a way to build a virtual network of contacts, but it requires a much larger effort, and its return isn't as immediately obvious. However, publishing articles gives us more recognition as an expert than just giving speeches. For people in the university and research environment, the papers typically need to be original research. For other people, the papers can be tutorials and opinions.

EXAMPLE: Attending professional meetings and conferences

While publishing papers and speaking at professional meetings are the best ways to establish visibility and networking, not everyone is in a position to do so. Simply attending professional meetings and conferences is the next best thing to building up a real network of professional contacts. It's also a good opportunity to build rapport with people we already know or just met. During most of such meetings and conferences, attendees have lots of free time when not in sessions, and it's common to see people chatting in the hall, going to lunch or dinner together, or sitting at a bar enjoying a couple of drinks. This is the time when rapport can be built and bonds established.

EXAMPLE: Delivering elevator pitches

Being able to give an elevator pitch at any moment affords us with opportunities to impress people in our network or to enlarge it with good contacts.

When we first meet people in an informal setting, we often only get a short window in which to impress them with our knowledge. Someone might casually ask us questions, such as, "What have you been doing?" or "What do you think of (a certain hot topic)?" Of course, they aren't interested in a detailed or long answer, so to impress them, always come prepared with several elevator pitches to quickly and intelligently respond to any spontaneous questions. I myself have lost several good opportunities to impress influential people just because I didn't have an elevator pitch ready. One of those occasions actually happened in an elevator! (See Chapter 1, Communications Smart, for more discussion on elevator pitches.)

A young IT professional told me once that his manager took him to an impromptu meeting called by a VP. During the meeting, the VP asked his manager a question about something related to what the young man's group was doing. Accordingly, the manager passed the question to the young man, but he was not prepared and could not give a concise answer. He lost an opportunity to impress both his boss and other senior managers at the meeting.

EXAMPLE: Enhancing rapport by being considerate

Obviously, rules such as keeping in touch with our network contacts and thanking them for their help are basic social etiquette, but we should extend these rules to include being considerate and sensitive to our contacts' time and feelings.

A professionally well-connected person, Andrew, was having a small open house party. In response to Andrew's RSVP request, a young man, Bob, responded positively but stated that he would be a half-hour late. Andrew told everyone at the beginning of the party that Bob would be there later, but Bob never showed up, which made Andrew feel slighted and unhappy. What Bob should have done was to show up with a nice little gift, even if he could stay for just a few minutes. Or, at the very least, he should have called prior to the party to apologize for being unable to go. The next day when Andrew and Bob were in the same committee meeting, Andrew picked apart Bob's presentation, a situation that Bob easily could have avoided.

Another professionally well-connected person, Cindy, CEO of a consulting firm, knew that Danny needed some advice. She sent an email offering her assistance and asked Danny to call her. Instead of calling, Danny emailed Cindy with a list of questions that would require only a few minutes' conversation on the phone but would be quite time-consuming to answer in writing, thus Cindy (being a busy person) never answered Danny's email.

Bob and Danny might never need help from Andrew or Cindy, and both Andrew and Cindy might have helped them anyway if Bob or Danny ever had the need, but it's always a good idea to maintain good rapport with the people we know, especially the influential ones.

Being considerate to someone's feelings and time might seem trivial, but sometimes the most trivial things can make a crucial difference.

EXAMPLE: Counterproductive networking

Networking is useful only if our contacts respect and like us. Otherwise, it can have a negative effect.

I knew of a salesman who could pick up a conversation with just about anyone, so naturally he knew a lot of people. Several companies, aware of his large network of contacts, hired him as a salesman, but what they didn't know was that because he had a reputation for being slightly wacky, his network contacts didn't want to do business with him. Any product he represented was viewed negatively, and he usually made his sales to those people who hadn't heard of him yet. In his case, knowing a large number of people not only didn't help him, it actually hurt him. He wasn't a successful salesman and rarely held a job longer than a year. The lesson here is that merely knowing many people isn't, by itself, considered "networking."

EXAMPLE: Importance of virtual networking: I

One day, I got a phone call from IBM telling me that it was hosting Mexico's Minister of Telecommunications, and the Minister wanted to use this opportunity to visit me. I was delighted and, of course, agreed. Apparently, one of the Minister's technical staff had heard a couple of my speeches and recommended the visit. Unbeknownst to me, this technical staff was part of my virtual network. (The same staff member had given me a bottle of tequila as a gift at the end of one of my presentations at a technical conference.) After this meeting, three events followed:

- Later, I visited the Minister in Mexico, where he agreed to be the honorary chair for an international conference that was to be hosted in Mexico City. (I was the conference chair.) To hold a conference outside the United States, we needed enough local engineers to attend the meeting in order to cover conference expenses. With the Minister of Telecommunications being the honorary chair, many local engineers were enticed to attend. The conference ended in the black. (Conferences are commonly in the red, particularly if they are held outside the United States.)
- I also facilitated contacts between the Ministry of Telecommunications and an office in the US National Institute of Standards and Technology to discuss possible cooperation.
- The request of the Minister's visit must have impressed people at IBM, who assumed that I had to be an expert in my field. Shortly after the visit, IBM approached me and offered me a consulting contract.

EXAMPLE: Importance of virtual networking: II

Virtual networking has played an important role in getting me the following two job offers and one consulting contract:

- When I applied for positions as Director of Computer Studies and tenured Full Professor of Electric Engineering and Computer Science, the recruitment committees wanted to go beyond the list of references I provided. One member of the committee knew Bob Lucky, who was a highly respected senior executive and a member of the National Academy of Engineering. Quite frankly, while I knew of him because of his achievements and his reputation, I did not expect him to know me. Surprisingly, he had apparently heard of me and gave me a very good recommendation. With that recommendation, the position was for me to accept or reject.

- Around 1990, Mitre Corporation was helping the federal government to review proposals that would provide a federal government-wide telecommunication service. One day, Mitre called and offered me a consulting contract. It turns out that somebody on the government side recommended me as a consultant. To this day, I have no idea who this person is.

- When I was interviewed for the position of Deputy Assistant Secretary for Information Systems at the US Treasury Department, the person interviewing me wanted to know how well known I was in the IT community. He asked the IT staff to look into it, and the response from the inquiry was very good. To this day, I have no idea who had provided positive recommendations to the inquiry.

RULE 6:   Being heard by listening

Be attentive to what others have to say: by doing so, we show respect to the speaker, learn the speaker's viewpoint, and are better prepared to respond.

Many of us tend to let our minds wander when others are talking. But there are situations in which we need to harness our wandering mind.

During a party, after we talk to one or a couple of individuals for a while, we may roll our eyes around to see who else is at the party. Our body language is showing that we are not being attentive. In a sense, we are slighting the individual(s) who are talking to us at the time. (Instead of showing any unpleasant body language, saying "I think I need to get another drink," and then slowly walking away is a more graceful way to move around a party.)

During a discussion with one or a few individuals, we may interrupt the one who is talking to inject our own comment or response. Not only is such an action impolite and irritating, but it may also prevent us from getting all the information we may need to get for a more informed response. (Instead of just interrupting, trying to get permission to interject by saying, "May I say something?" is a more polite way of interrupting. This assumes that we're confident we have all the information we need to make our comment.)

During a group discussion, we may be so consumed by thinking about what we want to say that we aren't paying enough attention to comments made by earlier speakers that could enhance our response. Consequently, we may not present our remarks as intelligently as possible. (A better approach is to write down key points being said by other speakers that could be incorporated in our own comments.)

🔳 EXAMPLE: Made a poor impression to a new boss

Shortly after his arrival, a new chief operating officer (COO) called for a staff meeting. Before the meeting formally started, there was a short social. The new COO worked the room in order to get to know his staff. When he was talking to Fred, a first-level manger, Fred was rolling his eyes to see who else had shown up and who had not. Obviously, Fred was not attentive to the new boss, and the new boss simply walked away. This surely was not a good way to start a relationship with the new boss.

🔳 EXAMPLE: The difference between "I heard you" and "I agree with you"

Tom, a second-level manager, was a very attentive listener. Whenever a staff member made a suggestion to him, he would listen attentively and appeared to be very much in agreement with the staff's opinion. To Tom, his attentiveness and agreeable manner meant only "I heard you," not "I agree with you," let alone that he was taking a suggestion. But almost every staff member walked way believed that Tom had agreed to, or would at least seriously consider, taking their suggestions.

Since Tom felt he had never promised that he would take any of the suggestions, none were implemented. From the staff's viewpoint, Tom had broken his promises. They were very mad, and even reflected their grievance to the director.

(Lesson learned: While we want to show attentiveness, we also want to be sure there aren't any misunderstandings as to what we have agreed to do.)

🔳 EXAMPLE: Gave an inappropriate response

The Representative of Taiwan was paying a courtesy call to North Carolina State University, which is located in North Carolina's Research Triangle area. The Chancellor, knowing that I am from Taiwan, asked me to attend his meeting with the Representative. During the meeting, I didn't pay attention to their conversation, and my mind wandered to something totally unrelated to the meeting. Then the Chancellor suddenly asked me, "Which locality has the most concentration of communication equipment manufactures?" Without hesitation, I said, "Richardson, Texas." From his facial expression, I could see that this wasn't what the Chancellor had wanted to hear. Apparently, he had been bragging that the Research Triangle and its vicinity had the largest concentration of communication companies. What I should have said was that the Triangle area had the most communication systems R&D facilities. (If I want to show off my knowledge, I could extend the answer by adding that Richardson has the highest concentration of communication equipment manufactures.) Had I paid attention to their conversation, I would not have made the mistake.

▣ EXAMPLE: Lost a consulting contract

Alan was submitting a consulting report to an engineering group at a large corporation. He joined the engineering group for a discussion about the problems the engineers had to deal with. Several members of the group made a presentation to highlight these issues. Alan was scheduled as the last presentation, detailing what he could do for them with a follow-up second-phase consulting contractor. While others were giving presentations and participating in discussions, Alan was busy fine-tuning his presentation and not paying much attention to other discussions in the meeting. When it was time for Alan to make a pitch for the follow-up contract, he looked like someone who didn't know what he was talking about. He repeated several problems that had already been addressed and solved by earlier presenters. As a consequence, he didn't win the follow-up contract.

CHAPTER THREE

Marketing Smart

Principle: Striking a chord with "customers"

Strategy: Promote our "product" (ourselves, our work, or our charge) in its best light

Almost everything we do in our professional life, whether we recognize it or not, is associated with "marketing." How successful our career is depends very much on the success of our marketing. The key to successful marketing is to present the "product" in a way that strikes a chord with "customers." This typically entices a positive spin on the product.

The word "product" is used here in a broad sense and includes any entity we want to promote. This can be ourselves, our work, a concept, a software, a hardware, and so on. The term "customers" is used here in a broad sense to mean anyone to whom we might need to deliver our products, including bosses, audiences, consumers, and clients.

Marketing is one form of "communication" in which we communicate to our customers the strength and merits of our products, with the objective being to rouse customers' interests and meet their needs. Prior to introducing a product, if we have influence in its development, we should incorporate features and functionalities that could resonate with customers.

Some of us, especially the technical ones, tend to belittle the concept of marketing. The argument is that if we're knowledgeable and accomplishing our tasks well, we don't need marketing. Fair enough. But believe it not, in reality, we do marketing all the time. We use oral reports, written reports, presentations, or even simple conversations to convince others, such as our bosses, about what

Fast-Tracking Your Career: Soft Skills for Engineering and IT Professionals, First Edition.
Wushow "Bill" Chou.
© 2013 The Institute of Electrical and Electronics Engineers, Inc. Published 2013 by John Wiley & Sons, Inc.

we know or have accomplished. These acts are forms of marketing. After all, the essence of *marketing is "convincing" someone of something*.

One reason that technical professionals tend to belittle the concept of marketing is their observation that some commercial salespeople use deception in promoting their products. My observation is that deceptive marketing strategies often backfire in the long run. It is important to remember that successful marketing needs to be based on facts, not deception or dishonesty.

Conventional wisdom tells us that we're doing a great job if we've successfully completed a task. Indeed, we have. But what really matters is how well stakeholders perceive what we've accomplished. In other words, the marketing skill is just as important as our actual accomplishment—if not more so. *The perceived success in what we've done relates directly to our marketing skills.* It therefore behooves us to be good in marketing—whether we're marketing our work, our products, or ourselves.

When we see people who are very successful in their careers or profession, we're likely looking at people who know how to successfully market themselves. This is most evident during political campaigns. Every candidate tries to strike a chord with a different audience using a differently slanted speech. Every time a political candidate is successfully elected, to a great extent, it is a direct consequence of smart marketing. (Here, the candidate is the product, and the voters are the customers.)

Regis McKenna, who played a crucial role in publicizing the Apple brand during the company's early years, was probably the first person to coin the phrase, *"Marketing is everything; everything is marketing."*[1]

Practicing the following rules, defined later in this chapter, can help us achieve deftness in being Marketing Smart:

- Sizing up and resonating with our "customers"
- Putting a positive spin on our "product"
- Making a convincing presentation with a well-crafted presentation
- Inciting enthusiasm with enthusiasm
- Being careful of careless comments (refer to Chapter 1, Communications Smart, for a general discussion of this rule)
- Using plain language (refer to Chapter 1, Communications Smart, for a general discussion of this rule).

In addition, the success of Marketing Smart can be enhanced by the following soft skills:

- Communications Smart (Chapter 1)
- People Smart (Chapter 2).

RULE 1: Sizing up and resonating with our "customers"

Design and/or present a product's features in a way that resonates with customers' needs, interests, and expectations.

As I have mentioned elsewhere in this book, people with different backgrounds and interests resonate differently. For example, some bosses are more interested in ROI (return on investment), while others are more interested in technical details. For exactly the same result from a task, we just need to emphasize and highlight the result differently depending on the boss's background and interests.

We would, of course, aim to have the contents of our product resonate with our customers. But the product's appearance and presentation are as important, if not more so, as the product's contents in rousing resonance with our customers. For example, suppose our boss asks us to make a presentation to a large audience. We can be sure that he would not be too happy if we make our presentation simply by using hand written transparencies on a projector. But we can be equally sure that our boss would resonate well if we gave a good dog-and-pony multimedia slide show of the same content.

The following story is a good illustration of the impact of resonance. (I got this from a viral email I received. Since I do not know who the original author is, I cannot give proper credit to its source. My apologies to the original author.)

A blind boy sat on the steps of a building with a hat by his feet. He held up a sign that said, "I am blind, please help." There were only a few coins in the hat. A man was walking by. He took a few coins from his pocket and dropped them into the hat. He then took the sign, turned it around, and wrote some words. He put the sign back so that everyone who walked by would see the new words. Soon the hat began to fill up. A lot more people were giving money to the blind boy. That afternoon, the man who had changed the sign came to see how things were going. The boy recognized his footstep and asked, "Were you the one who changed my sign this morning? What did you write?" The man said, "I only wrote the truth. I said what you said but in a different way." What he had written was, "Today is a beautiful day and I cannot see it."

EXAMPLE: Resonating with division heads to move "hobby shops" under a CIO's control

This is about a CIO's successful strategy of integrating divisions' "hobby shops" into his office by making a proposal that resonated with division heads' interests.

At Corporation I, the IT shop is headed by a CIO and provides IT services to the entire organization. Included in these services is the development of application software for internal use. Under the CIO, a CTO was charged with overseeing this task, but historically, the CTO office wasn't able to satisfy the needs of many divisions in the company, so several "hobby shops" popped up throughout the organization. (A hobby shop is a small IT group that develops an IT application, typically a software system, for use in its own division. The work is done outside the realm of the CIO's authority. Examples include applications for multimedia, e-commerce, staff training, personnel management, and client management).

When a new CIO came on board, he wanted to squelch these hobbyshop activities and make the CTO office to be the sole service provider for application software development, or somehow integrate the hobby shops into the CTO office. He could have just enforced the charter of the CIO office, but this might have incited political infighting and upset people. Instead, he took a mild marketing approach. He visited all divisions with hobby shops to better understand their functions and capabilities as well as their inadequacies.

A month later, he went back for a second visit. He told each division head that the CTO office had strong expertise and substantial resources. If the division would let the CIO and CTO offices take responsibility for its IT application needs, he promised specific enhancements and additional features as well as better maintenance and prompt support. He also said he'd shorten the development time for future requirements. Finally, he said he'd do all of this for less cost. He then suggested structuring this service arrangement on a trial basis. After 1 year, if the division wasn't pleased with the CTO office's service, it could take back responsibility of the IT application.

The CIO did not win over everyone, but many accepted his offer. The CIO's proposals resonated well with the divisions on performance and cost savings, with no risk to boot.

EXAMPLE: Resonating with the boss

This is about a company that wasn't very computer savvy. Senior management viewed the computer and communications services as glorified utilities and saw the CIO as simply a utility manager. Of course, the CIO thought otherwise. He felt that there was much more his office could do to support the company's mission.

The company did not do business with the general public, so senior management did not give much priority to the company's website. Again, the CIO thought otherwise.

The CIO knew that he needed to convince the CEO of IT's importance to the company's future to obtain additional funding and support. He decided that adding a "resonating" feature to the website would be the simplest way to get the CEO's attention.

He learned that the CEO was an ardent fly-fishing sportsman, so he had the webmaster create a fly-fishing icon. Clicking the icon, an impressively formatted table appeared with various fly-fishing websites and their specialties.

The CIO requested a meeting with the CEO to discuss the budget. At the meeting, he casually showed the sample website with the fly-fishing icon to the CEO. The icon and its functionality caught the CEO's attention and resonated with his interests. At the end of the meeting, the CEO promised additional funding to strengthen the company's Web appearance, which was a first step toward using IT to better support the company's mission.

RULE 2: Putting a positive spin on our "product"

Put a positive spin on the strengths and weaknesses of the product to place it in its best light.

Whether a deliverable is a software, hardware, presentation, completed task, or personal skill, we need to emphasize its strength and its current and potential benefits, as well as downplay or gloss over potential weaknesses. These positive spins place the product in its best light and will enhance its chances to resonate with customers.

> **EXAMPLE: A positive spin in marketing a software system**
>
> A major part of X Corporation's business was consulting with various organizations to analyze and configure their Internet connections. X Corporation's engineers thus developed several ad hoc network design software programs to help them with their analysis. X Corporation's marketing director thought it would also be a great idea to offer these software programs to customers as a subscription-based cloud computing type of service.
>
> However, there was one big problem: the software programs were developed independently. They didn't even have common input/output (I/O) formats. So, X Corporation's engineers developed a front end that provided an integrated common access point for modular network design software programs.
>
> With a positive spin, X Corporation named the combined software system "MIND" for Modular Integrated Network Design. A very impressive name!! X Corporation turned weakness into strength. (As MIND evolved, new features were added and old functions were improved or removed. The revised product brought in good revenue for the company for many years.)

RULE 3: Making a convincing presentation with a well-crafted presentation

Craft presentations to convey our positive spin and deliver information that can resonate with our customers.

A presentation is where the rubber meets the road. The presentation may assume various forms—oral, written, informal conversation, flyer, dog-and-pony show, or advertisement. A good presentation can make a poor product look good and vice versa.

In our professional life, we frequently need to deliver oral presentations to a group of people of varying sizes. For most of us, the easiest, most common, and often necessary aid used to craft our presentation is a set of slides. The slides should concentrate on key points and key information, use words and phrases understandable by our listeners, and apply color highlights to help the audience grasp the information we want to convey. The slides may not need to be dressed up if the presentation is given to colleagues working in the same office, but they may need to be dressed up as a multimedia show if given to a large audience in a large conference room. The objective of the dressing-up is to project an authoritative image of the presenter and/or enhance the presentation's resonance with the audience. The dressing-up should not distract from the effectiveness of the presentation.

▌ EXAMPLE: A simple crafted presentation leading to a contract

In the early 1980s, I frequently gave speeches at professional conferences on computer communications (precursor of the Internet). At the end of one such session, the conference organizer, Gene, approached me and asked whether I would be interested in conducting a tutorial when the same conference was to be held the following year, with a promise of a substantial amount of remuneration.

Gene then told me he was at my session. (Because of the large number of attendees, I did not notice his presence.) He confided that he particularly liked the transparencies I used for my presentation. He said they had helped make my lecture very understandable, even enjoyable. This was a bit of a surprise: most of the other speakers worked for large corporations and used professionally prepared and machine-produced transparencies. At that time, I was a university professor and did not have that luxury. All my transparencies had to be written by hand, without any support from draftsmen. But it turned out that my weakness became a blessing: the others needed to submit their material to the transparency production staff many days ahead of the scheduled talk; I prepared my presentation and transparencies on the day before my trip to the conference. I had the advantage that the key points in my transparencies matched well with the flow of my speeches. The others' transparencies were all in black and white. (At that time, all machine-produced transparencies and slides were black and white.) Mine were in color. (I used pens with various colors when preparing these transparencies.) Looking at the colored transparencies was not as boring as viewing the black-and-white ones. I could use different colors for different sets of key points. I could use color to place emphasis on certain aspects of the transparencies. (I had another advantage, too—I did not personally write/draw my transparencies. My wife did. She has beautiful handwriting, especially in block letters. Regardless of color, her handwriting looks less boring than machine-generated block letters.)

RULE 4: Inciting enthusiasm with enthusiasm

Present the product with outward excitement, confidence, and/or enthusiasm. Enthusiasm is contagious. Our enthusiasm can incite enthusiasm in our customers.

Displaying enthusiasm is important for marketing any product. We can find good role models in our elected officials. A common trait of all speeches given by successful elected officials is their exhibition of enthusiasm and confidence. This is how they inspire their customers (the voters).

Those of us who have attended conferences with parallel sessions and multiple speakers probably have observed the following: an enthusiastic speaker typically draws a considerably larger audience.

Whether making a presentation of a report or proposal, our enthusiasm can help us get a positive response from our customers.

The most well-known role model for giving crafted and enthusiastic presentations was Steve Jobs. But I believe Leonard Kleinrock is an even more impressive showman/presenter. Anyone interested in being a greater presenter should watch and study their presentation videos.

A MARKETING ROLE MODEL: STEVE JOBS (AND HIS EMBODIMENT, APPLE)

Resonance

In developing its products, Apple places utmost importance on resonating with its customers' interest. Its designers are talented at projecting what customers want and desire. Apple leads its competitors hands down with "user friendliness" and "sleek design" (but not necessarily with inventing innovative technology). We all know of people who converted to Mac because they were frustrated with craplets and crapware preinstalled by PC vendors. We also know of people lured to Apple because its products have that sleek, gee-whiz appearance. It's as though users designed the Apple products, while competitors' products were designed by and for geeks and bean counters.

What has always amazed me is why PC and other digital device vendors cannot seem to make their products as sleek as those of Apple's. (One possible exception is Sony.) During a conversation with a product manager of a large PC manufacturer in 2010, I asked him this question. He proudly told me that his company's new products now have a sleek appearance. Poor fellow—he just doesn't get it. Compared with its last generation's products, yes, the newer products indeed are sleeker. But they still aren't as sleek as the products from Apple.

With Apple's continued success in its iPod, iPad, iPhone, and MacBook Air product lines, these other vendors seem to have finally woken up to the importance of sleek design, but they cannot create their own designs, they just imitate Apple's. (As of this writing, early 2013, its competitors have "leaked" out information about their conceptual tablets and smart phones that look sleeker and have better functionality than those of Apple. But I don't know when they will be able to convert the concept to actual products. Besides, by then, Apple's products may also be sleeker and have more enhanced functionalities.)

Positive spin

Of course, no product is perfect, but Apple is very good at emphasizing its products' strengths and glossing over their weaknesses. For example, when Steve Jobs first announced the iPhone, he promoted its new features and potential, with scant mention of its "telephone" features. It was, after all, foremost a telephone—just not a very good one. It had mediocre voice quality and a meager battery life. Yet with Jobs' successful spin, few customers noticed this deficiency or even cared.

Similarly, when Jobs first presented the iPad, he touted it as a "netbook" that could be connected to peripheral devices. He cleverly glossed over the fact that the iPad did not have a USB or fire wire port for directly connecting to such devices. (A dock station is needed for such connection.)

Crafted presentation

Apple is very clever in presenting its new products. Before announcing a new product, it leaks a few details to arouse potential customers' expectations. Then, when making an official product announcement, Steve Jobs would perform a good dog-and-pony show to resonate with his audience about the new product's wonderful features. As one of the world's most polished presenters, Jobs was very skilled at putting a positive spin on Apple's products. Product presentations were further enhanced by a superlative website and advertisements. (Now that Jobs is gone, we will see whether the new CEO at Apple can fill his shoes.)

Enthusiasm

When introducing new Apple products, Steve Jobs always exuded exuberance, confidence, and optimism. His enthusiasm was quite infectious to his audience. When he introduced Apple's first-generation iPad, he called it "a truly *revolutionary* and *magical* product." Not everyone was sold by his proclamation, but plenty of people camped out to buy the product.

While I would not call the iPad a *magical* product, its introduction did kick off a device revolution. But this was not because the iPad itself is revolutionary: it was Apple's and Job's mastery in marketing that induced the revolution. Jobs' enthusiastic introduction of the iPad caused a tsunami of demand for the iPad. Competitors took notice and rushed out similar tablets. Everybody and their uncle dreamed to develop best-selling app(s) for the new device. The combined effect was indeed revolutionary.

Apple and Steve Jobs' iconic marketing skills in resonance, positive spin, and crafted and enthusiastic presentations made Apple the most valuable company in the world and revolutionized handheld digital devices.

REFERENCE

1. McKenna, R., "Marketing Is Everything," *Harvard Business Review* 69(1) (1991): 65–79.

PART THREE

Dealing with the Self
THE BASIC

CHAPTER FOUR

Work Smart

Principle: Focusing on the best return with reasonable effort

Strategy: Seek out methodologies that can bring optimal returns from our efforts; reject, give up, or adjust tasks that do not otherwise net good returns

In 1970, ARPA (Advance Research Projects Agency) commissioned a mathematic optimization expert, Howard Frank, now a member of National Academy of Engineering, and his team to configure its incipient ARPNET, a precursor of Internet. The effort involved the selection of switch sites, the placement of transmission lines, and the selection of transmission speeds. The object was to determine a least-cost configuration that would allow the network to handle traffic while meeting performance requirements. It was immediately determined that a true optimum solution in a mathematical sense would require a tremendous amount of effort, if possible at all, and its benefit could not justify the efforts. The team developed approximating heuristic algorithms to find the best solution (from the pragmatic sense) with a reasonable amount of effort. *Their strategy: develop a methodology that can bring best returns with reasonable efforts.* (Since then, these algorithms have frequently been used by others to configure Internets.)

Larry Page, a cofounder of Google, assumed the position of CEO in April 2011. His first order of business was to drop more than 25 ongoing projects at Google. The company made the following statement as the reason for doing so: "We're in the process of shutting down a number of products which haven't had the impact we'd hoped for." In other words, Google did not expect these products to bring enough return for the effort going into them. Therefore, it was refraining from investing additional resources into these projects. *Its strategy: give up products unlikely to give returns as good as other products.*

Fast-Tracking Your Career: Soft Skills for Engineering and IT Professionals, First Edition. Wushow "Bill" Chou.
© 2013 The Institute of Electrical and Electronics Engineers, Inc. Published 2013 by John Wiley & Sons, Inc.

Practicing the following rules, defined later in this chapter, can help us achieve deftness in being Working Smart:

- Achieving outstanding results by not seeking perfection
- Avoiding blunders of overconfidence
- Focusing on self-examination, not on blaming others, when things go awry
- Killing two birds with one stone (refer to Chapter 5, Time Smart, for a general discussion of this rule)
- Never polishing a sneaker (refer to Chapter 6, Career Smart, for a general discussion of this rule).

RULE 1:   Achieving outstanding results by not seeking perfection

Refrain from being a perfectionist; instead, aim to achieve the best return on investment (ROI). A zeal for perfection tends to draw more resources, but ends up with a lesser result.

A pragmatically excellent result—say, between 70% and 90% of a theoretically perfect solution—can often be obtained with a reasonable amount of effort. To go further than this typically requires a substantial amount of additional time and effort, but such added effort and time often cannot justify the relatively small amount of marginal return. Furthermore, it may even do more harm than good: it might lead us to lose sight of the problem, and hinder or distract us from pursuing other opportunities. The irony is that the "perfect" solution might not even be correctly definable.

Figure 4.1 illustrates graphically the relationship between the amount of return and the amount of effort or time needed. It approximates a negative

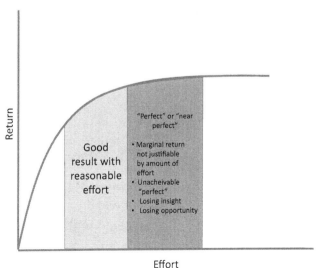

FIGURE 4.1   Good versus perfect

exponential distribution. The light gray portion represents good results that can be achieved with a reasonable amount of efforts. The dark gray represents "theoretically perfect" or "near perfect" results which are only marginally better than the good results but require a substantial amount of additional efforts.

In one of his speeches, President Barack Obama said, "We can't afford to make perfect the enemy of the absolutely necessary."

EXAMPLE: Perfect beginnings, mediocre endings

Many projects end in mediocrity because a significant amount of effort was put into the project at the beginning and there was not enough time left at the end to get it done properly.

We've all witnessed the following scenario at some point in our lives. At the beginning of a project, the people involved are very enthusiastic and strive to do a perfect job. But they spend so much time aiming for perfection that by the time the project is about to be due, a substantial amount of it is still not done. The participants have no choice but to do a rush job just to finish it. If they were not so ambitious initially, they could have done a good job instead of ending up with a mediocre one.

I have seen many students fail to turn in project assignments, not because they didn't put in the effort but because they spent too much energy aiming for perfection at the beginning and did not have enough time to finish the assignment at the end.

(*Caveat*: There is a big difference in putting substantial front-end effort between trying to be perfect without proper planning and trying to get a good ROI with proper project planning. The former leads to mediocre results while the latter leads to good results.)

EXAMPLE: Lost opportunities while trying to be perfect

Setting an unrealistically "perfect" goal not only leads to a likely unsatisfactory result, or even failure, but also contributes to missed opportunities.

A smart technical manager at a large company was given a large amount of resources to develop tools for Internet configurations. He decided that with computing power so cheap now, he could develop a mathematic programming software program to get a perfect solution for configuring an "optimal" Internet connection that would satisfy all traffic and performance requirements. (The tools available for such purposes are all of a heuristic nature, meaning they'll give good answers but not theoretically optimal ones.)

However, to get a perfect solution, the program needs perfect inputs, and in this case, one of the inputs is the projection of the traffic volumes, rates, and patterns of various data messages being sent through the network. Because there's no way to give a perfect projection, a perfect solution cannot exist. Furthermore, the computational complexity involved is extremely large, so it would take an unreasonably long time just to get one answer.

Thus, the software he developed had only limited pragmatic value. Had he used the same amount of money and time to work on something more pragmatic, he could have developed far more useful tools to show to his company management. He could also have been better recognized professionally. By all accounts, this was a lost opportunity.

RULE 2:   Avoiding blunders of overconfidence

Be wary of stepping from confidence into overconfidence. Overconfidence tends to underestimate the amount of resources and efforts needed for success and ultimately ends in lesser results, if not total failure.

Yes, it's important to exude confidence, but there's a thin line between confidence and overconfidence. While the former is crucial to our success, the latter can have the opposite effect. Overconfidence could result in many "bad" consequences:

- Underestimating the resource and efforts needed for completing projects
- Making decisions without the benefit of advice from others
- Overestimating our value and importance
- Holding an attitude and mannerisms that are offensive to others
- Muddying our analytic ability.

EXAMPLE: Hindering career advancement

Many highly competent technical professionals are so confident in their technical supremacy that they believe they can advance in their career without paying attention to soft skills or managerial aptitude. They end up frustrated.

Andrew was the most competent technical staff in his group. He was certain that he would be in line for promotion to manager. But when the manager position became vacant in his group, Bruce was promoted to the role, not Andrew. Andrew was surprised and upset because his technical skill was definitely superior to that of Bruce's. Andrew's problem was that he was proud of and overconfident in his technical skill. He did not appreciate the need for managerial skills and spent no effort familiarizing himself with these skills.

EXAMPLE: Project overrun

Most engineers, including computer engineers, are quite smart but tend to be overconfident of their capabilities. They are prone to underestimate the time and effort needed to complete a project, be it a software system or a bridge construction. This is why almost all large projects end up needing more resources and time to complete than originally estimated when the projects commenced. When I was a technical manager, I always added 20% on top of the required time for completion as estimated by technical staff.

■ EXAMPLE: Losing a presidency

When people are overconfident, they tend to be careless in their analysis and sloppy in their behavior and verbal expression. Do you remember the first US presidential debate in 2000 between Al Gore and George W. Bush? Gore was obviously overconfident, and it showed in his mannerisms—frequent sighing, rolling his eyes, and so forth. Needless to say, this didn't go over well with some voters and might have adversely affected his vote tallies in closely contested states.

■ EXAMPLE: Losing an investment opportunity

Charles developed a software package that roused the interest of several venture capitalists. But due to a combination of his lack of business acumen and his overconfidence, he was quite rigid in his demands. As a consequence, he and the venture capitalists couldn't close their differences on issues of royalties, compensation, and ownership. If his overconfidence hadn't muddied his judgment, he could have made a deal with one of the venture capitalists and been a mega-millionaire by now.

■ EXAMPLE: Losing the opportunity to be a valedictorian

Kent was a straight-A student all his life. During his college senior year, he took a course with a prerequisite that he hadn't taken yet. By midterm, he had a B. To graduate as a valedictorian, he needed to get an A in very course he took. Since it was not a required course for graduation, he should have dropped this particular course to ensure his chance of being listed as a valedictorian, but he didn't. He was confident (actually, overconfident) and believed he could turn the B into an A by the end of the semester. He didn't—and it was the only B he received in his 4 years as an undergraduate. He could have been a valedictorian had he not been so overconfident and dropped the course.

RULE 3:	Focusing on self-examination, not on blaming others, when things gone awry

Focus on ourselves in mapping out a corrective path; pointing fingers at others when things gone awry does not in general help matters and can even be counterproductive.

Confucius said, "When an outcome is not as good as you have expected, count on yourself to find a right solution to correct it." We all tend to blame others when things do not turn out the way we had hoped. But ultimately, it isn't really important who's at fault—instead, we should focus on what we could have done differently or what we should do differently in the future to prevent the same thing from happening. We cannot control others, but we surely can control ourselves. Pointing fingers at others does not help matters.

EXAMPLE: Workplace culture

It is our responsibility to integrate ourselves into our workplace's culture. We should not expect our workplace to adjust to ours.

A senior director of a manufacturing company, Ken, complained that because of his ethnic background, he would never be promoted to a VP position. Upon further questioning, I learned that he lived in California and worked in Washington, going home to California every Friday afternoon and returning to work in Washington every Monday morning. I then asked him how he socialized with corporate management to establish a good rapport with them, and he said he couldn't because of his lengthy commute. Therein was the problem—if he didn't want to spend the effort to mingle with corporate management, how could he expect corporate management to accept him as one of them? After some reflection, he admitted that if he had stayed in Washington from time to time instead of going home to California every weekend, he might have a decent chance getting promoted. He could not blame corporate management for not getting a promotion.

EXAMPLE: Chronic complainers

Chronic complainers are losers! They tend to fault others for every one of their failures and aren't willing to take personal responsibility to improve their chances of success. Unfortunately for these people, they have a warped logic, and unless they adjust it, they're likely to have only limited career success. We all have known and observed people like this.

Being a teacher for many years, I've seen plenty of examples. Once, a student sent me a very bitter email complaining about the amount of material to be covered in the midterm: "I have to work 40 hours a week. I can't spend a full day to prepare for the exam." It was admirable that he took classes while working full time, but he should have been prepared to put enough time into the classes he signed up for. No one can expect to get a passing grade without putting enough effort into learning the required material, regardless of the reason.

CHAPTER FIVE

Time Smart

Principle: Recognizing time as more valuable than money

Strategy: Aim for good return on investment (ROI); turn spare time into opportunities; minimize nonproductive time

In his bestselling book, *The Last Lecture*, Randy Pausch told the following story. At a grocery store checkout scan machine, he was overcharged by about $16. He estimated that it would take 20 minutes for him to go through the process to get the money refunded. He decided not to seek the refund. He felt he could spend the 20 minutes on something more meaningful than getting the $16 back. Randy Pausch was trying to point out that often, time is more valuable than money.

We all have exactly the same limited amount of time in a day. But the quality and return in spending and investing time can vary. The time is of good quality if we spend and invest it with a clear and focused mind. The time offers good returns if we spend and invest it by seeking to maximize its return in value.

Investing and spending time is analogous to investing and spending money. When we invest money, we want to maximize our ROI (return on investment). When we invest time, we want to maximize our ROI on that as well.

Strictly speaking, of course, ROI in time is not the same as ROI in money. ROI in money is expected to return more in money or a monetary equivalent. ROI in time cannot be expected to return more in time; instead, the return is expressed in terms of number and quality of successfully completed tasks and/ or the amount of financial return.

I have had some interactions with a couple of venture capitalists, all of them are very keen about ROIs and are particularly cognizant of the value of time. A

Fast-Tracking Your Career: Soft Skills for Engineering and IT Professionals, First Edition.
Wushow "Bill" Chou.

few years back, two venture capitalists from California visited me on the east coast independently on different days. Interestingly, they acted similarly. They both flew in late in the afternoon, stayed at an airport hotel, asked me to meet them at the hotel, and flew out early the next morning. I do not think these coincidences were by chance. They both wanted to minimize nonproductive time. During our conversations, one of them said bluntly, "I have money, but I do not have time. I want a prompt response. I do not want our discussion to be drag out for a long time." What he was saying was "time is more valuable than money" and timewise, he was only interested in "good ROI."

Practicing the following rules, defined later in the chapter, can help us achieve deftness in being Time Smart:

■ Investing time with the same zeal as venture capitalists investing in money

■ Killing two birds with one stone

■ Minding ROI

■ Making nonproductive time productive

■ Turning spare time into opportunities

■ Keeping the mind sharp by taking catnaps.

RULE 1: Investing time with the same zeal as venture capitalists investing money

Invest intelligently and sufficiently at the onset, with the objective of gaining a huge ROI at the end.

A venture capitalist (VC) typically does due diligence in searching for appropriate projects in which to invest. Once (s)he identifies one, (s)he would take a calculated risk and commit a sufficient amount of resource at the onset with what (s)he believes is needed to maximize the project's chance for success. (S)he usually puts little additional resources, if at all, during the intervening period.

With similar zeal, we should, at the outset of a complicated task, make a sufficient amount of investment (in time, planning, and effort) to ensure the task's successful conclusion with relatively limited additional follow-up effort from ourselves.

Technical professionals actually have one big advantage over VCs: techies understand the projects they work on much better than VCs understand theirs. While the success rate for VCs is low, the success rate for technical projects can be very high. (Note: for VC investment, very high ROI for successful projects compensates for all that is lost in the failed ones.)

📖 **EXAMPLE:** The wisdom from the Facebook phenomenon

In the movie *The Social Network*, Facebook's Mark Zuckerberg and Napster's Shawn Parker both come to the same conclusion in Facebook's early years: instead of cashing in for tens of millions of dollars, more time and effort should be invested in Facebook to make it "cool," which would make it worth billions of dollars. We cannot even dream of being an accidental billionaire. But the underlying principle of this argument is valuable to all of us: investing sufficient time and effort intelligently at the onset and resisting the temptation to cash in for a quick return could bring huge returns, that is, huge ROI, in the long run.

I know people who have had excellent ideas for software systems but were not able to devote enough time at the onset, lacked the zeal of a venture capitalist, got side-tracked by a limited immediate return, and consequently failed in their original more ambitious endeavors.

📖 **EXAMPLE:** Juggling multiple jobs

Kirk is the epitome of a person who knows how to invest his time productively, just like how a VC invests money. He's at once a renowned researcher, a chaired professor at a top university, and the owner of three companies.

As a researcher, he puts in time and effort in writing excellent grant proposals. Once a grant is awarded, he spends time working out a good research plan. He then delegates the bulk of the work to his research assistants. When the grant report is due, he summarizes the research results from his assistants and presents an excellent report to the research sponsor.

As a professor, Kirk also puts time and effort at the onset and does a great job preparing his lectures and class notes. He then modifies his lectures to conduct numerous off-campus lectures and short courses, and he has published his notes as popular textbooks.

As an entrepreneur, using his technical expertise, he began with a small-scale high-tech company and planned out the company's strategy. After his initial efforts in getting the company started, he hired a COO to actually run the company and expand the business, with a relatively small amount of his own effort.

By investing time intelligently and sufficiently at the outset of each task he undertakes, Kirk has been able to perform each task successfully (with a minimal amount of his own time) and harvest huge returns.

📖 **EXAMPLE:** Developing a software system

In developing a software system, if we map out a detailed, well-planned specification, we are bound to spend more time at the project's outset. If we then follow and enforce the rules defined in the specification, we can save substantial amounts of time in the long run. The system can be maintained easier and need fewer adjustments. We won't need to change as many requirements or redo configurations for different applications.

Unfortunately, more often than not, either people don't have a well-planned specification (not willing to put enough effort at the onset) or, if they do, don't follow through on it and take shortcuts (take a short-term quick return, instead of a long-term better return).

(*Caveat*: There is a big difference in putting substantial front-end effort between trying to be perfect without proper planning and trying to get a good ROI with proper project planning. The former leads to mediocre results while the latter leads to good results.)

EXAMPLE: Running a conference

Some conference organizers complain that running a conference demands too much time and effort, yet others don't. This difference in opinion hinges on how well the conference is planned at the outset.

If the organizer (mostly like the general chair) spends time and effort at the very onset by laying out a good plan, forming a capable committee and subcommittees, and delegating responsibilities to various committee members and subcommittee chairs, a very successful conference can be carried out with little need to spend much time personally in managing the details. (Otherwise, not only the effort can be quite daunting, the conference may not turn out well.)

RULE 2:   Killing two birds with one stone

When feasible, plan to make one deliverable useful for multiple tasks.

With proper foresight and planning, the same deliverable for one task can be modified and used for other tasks. This is a great strategy to multiply our productivity, save time, improve ROI, enhance performance, and increase our visibility.

Many people seem to believe that what they do is very unique and they cannot make use of this one-stone-two-birds strategy. I contend that this is not always true. If we look hard, we will find this strategy can be applicable to some of what we are doing—not all of the time, but some of the time.

Looking for a second or third bird to hit with the same stone that is intended to kill or has killed the first bird has been a credo in my professional life. The person who instilled this strategy in me is Howard Frank, a member of National Academy of Engineering. He was my thesis advisor, onetime boss, and mentor. I learned from him this two-bird, one-stone strategy, and it has worked out for both of us.

EXAMPLE: Oral presentations

Some of us have made frequent oral presentations within our own workgroup. Most of these aren't of much interest to people outside the group. However, when we do make a presentation that might be of interest to others, we can easily make similar presentations outside the group, outside the organization, or even at conferences. "Killing" a second or third bird with the same stone—that is, modifying an internal presentation for outside consumption—still takes time, but this additional marginal investment in time can enhance our work's visibility and build our reputation as an expert.

EXAMPLE: Written reports

Just like oral presentations, we can similarly publish internal reports of interests to others, such as at a conference or in a magazine/journal. And just like
oral presentations, modifying an internal report for outside consumption still
takes time, but this additional marginal investment in time can enhance our
reputation as an expert.

A few years back, a book editor was planning to publish a book with a
collection of tutorial articles. He approached Helen and Al, asking each to
write a chapter relating to their expertise. In essence, the writing effort would
involve extracting, assembling, and modifying some of their internal reports
that were available to the public. Helen thanked the editor profusely for the
opportunity, while Al declined, saying he wouldn't do it because his boss
didn't view writing an article as a job assignment, and he did not want to
spend his own time writing the article. In the years to follow, Helen continued to look for presentation and publishing opportunities; Al did not.

Today, Helen, who has an MS from an average university, is a very successful senior executive and an IEEE fellow; while Al, who holds a PhD from
an Ivy League university, is still a staff member. No doubt, Helen's success
depends far more than her one-stone-two-birds skill she holds. But it has
surely helped.

EXAMPLE: Consulting

A consultant will sometimes get a contract for a project despite knowing
little about the topic. He wins the contract by initially acting knowledgeably
about the project and then, after winning the contract, spends time learning
about and developing tools for it.

At the end of the contract, the consultant may have made very little
profit because the substantial amount of effort he put in to master the
project. But he has learned a great deal. He can then present himself as an
expert on the subject matter, actively looking for a new client with similar
problems. He uses essentially the same tools he developed for the first client.
Quite a few consultants have made good profits using this strategy of "killing
multiple birds with one stone."

EXAMPLE: Using form letters

We've all received form letters. Some of us have even sent them. Isn't it nice
when they're more personalized? With proper planning, this is often easy
to do.

One of Shawn's responsibilities was to respond to customers' inquiries.
He spent a substantial amount of time analyzing all possible scenarios and
designed 12 form letters. He also marked where to insert information specific
to an individual's inquiry.

He gave the form letters to his secretary and showed her how to individualize them. She would bring an inquiry to his attention only if it didn't fit into any of the previously identified scenarios. So, even though Shawn saw very few inquiries himself, every response appeared to be coming from him personally. The relatively small amount of time spent on planning at the outset resulted in a substantial amount of time saved. One stone killed many birds.

RULE 3:   Minding ROI

Given a choice, opt for the task with better ROI; given an opportunity, enhance a task's ROI.

We routinely encounter tasks that generate low ROI, and we routinely carry them out anyway, which is akin to wasting time. We need to avoid, adjust, or find alternatives for these tasks.

When given an option in choosing our tasks, we should select the one with the potential of the best ROI in time. For whatever we choose or are required to do, we should always strive to get the best results with minimal time.

EXAMPLE: Letting others do copyediting

Both George and Sam had a PhD, worked at the same high-tech company, and shared the same secretary who had a degree in English and copyedited their writings. This did not please George. He openly said, "I am a PhD and don't need anyone to correct my writing."

On the other hand, Sam was delighted. This situation allowed him to concentrate on the contents of his reports; the secretary could do the polishing. He felt he got a better ROI by concentrating on the contents without having to worry about grammar and style, and the time saved from delegating copyediting to the secretary would free him to work on other tasks. Maybe it was just coincidence, but Sam ended up with more publications and achieved a higher professional status than George.

EXAMPLE: Getting technical expertise from friends

With the technology expanding so rapidly, we frequently need to learn new technical information. We can always read books/publications or search the Web for the needed information, but that's time consuming and doesn't always provide the exact information we are looking for. An alternative that is far easier and faster is to pick the brains of knowledgeable friends.

A few years back, I was working on a consulting contract and needed some expert knowledge on building wirings. A friend of mine was an expert on the topic. I prepared a set of questions and called him. I learned in 30 minutes what would otherwise have taken tens of hours for me to figure out.

EXAMPLE: Refraining from attending inconsequential meetings

We've all gone to meetings and conferences to listen to others' speeches. How often is the return from attending such conferences and meetings worth the time spent?

We listen to a speech for different reasons: entertainment, inspiration, special occasion, obligatory duty, and information. Any of the first four reasons might justify the time spent but not necessarily the last one. (If the information is of interest to us, but not available in print or on the Web, it might justify the time spent to attend the speech. Otherwise, no.)

If we wanted to attend a Steve Jobs speech for entertainment and inspiration, by all means, we should have gone. But if our purpose was purely to learn what he was going to announce, we could have spent much less time getting the same information. Almost immediately after his talk, the speech was posted on Apple's website, and a summary of its key points would be available on other sites, freeing up our time for more productive tasks.

EXAMPLE: Beware of perfection

The worst return on time investment is probably trying to be perfect. As the output of a task approaches perfection, it takes a substantial amount of time for just marginal improvement—and it's difficult to define "perfect" anyway.

Consider a tightly written computer program. Some programmers are proud of being able to write concise computer programs. They spend more time than necessary in finishing the program. Yet a concise computer program isn't a perfect or even a good computer program; in fact, it could be a very poor one. The computer program can end up being more difficult to maintain and enhance.

RULE 4: Making nonproductive time productive

Do something productive when attending unavoidable, not fully productive activities—but only when feasible and appropriate.

As necessity dictates, we often have to spend time on activities that aren't very productive, such as attending obligatory meetings. For these activities, while we must be physically present, we are not fully engaged mentally. Being *creative* and *judicious*, we can potentially make some of these nonproductive times more productive by doing one of the following: thinking, reading, doing a breathing exercise, or working on smartphone/tablet/laptop devices.

(*Caveat*: Before we engage in such activities, we need to assess the situation to ensure that the level of engagement would be appropriate and/or not offensive.)

On the other side of the coin, we may be fully engaged mentally but not physically. We can use this opportunity to do some limited physical exercise, such as stationary biking.

🖳 EXAMPLE: Business/social meetings at lunches or conferences

Many people, especially senior managers, use lunchtime for business meetings or as an opportunity to build rapport with colleagues and professional friends. When attending conferences, many people use the cocktail hour and dinners for the same purpose.

🖳 EXAMPLE: Working while commuting

We all spend a great deal of time in commuting. How well we can harness this seemingly wasted time could impact significantly on our career. Some use the commuting time to think, such as solving a technical problem or developing a business strategy; others use the time to listen, such as a tutorial based on a CD. (*Caveat*: If we are driving, our concentration must be on driving. Thinking or listening should not in any way interfere with or distract from our driving.) If we are passengers in a vehicle during our commute, we can work on a laptop, read, or even conduct business. I know a CEO of a large company who routinely asks his staff to ride with him so that they can brief him on various upcoming events.

🖳 EXAMPLE: Exercising while commuting/working

I have known people who use commuting as an opportunity for exercise by jogging or biking to work. (Of course, do take a shower immediately after jogging or biking. The appearance and smell of heavy sweating could be offensive to many people. I know of a bike-riding assistant professor who was given his promotion on the condition that he must take a shower immediately upon his arrival at the university. Students had been complaining about his foul odor!)

Some people set up a stationary bike in their offices and ride it while reading, thinking, or even conversing with visitors. Handy exercise equipments, such as dumb bells and grip exercisers, are also widely used.

🖳 EXAMPLE: Multitasking at boring meetings

From time to time, we are obligated to go to meetings where we are marginally involved. Depending on the nature and format of these meetings, we can conceivably multitask on reading, thinking, writing, or working on our smartphone, tablet, or laptop. (Please see the "Caveat" statement appearing at the beginning of this section.)

I have observed at several professional committee meetings where working on a laptop is acceptable that there are always a couple of individuals diligently typing away on their laptops. The interesting, yet not surprising, part of my observation is that each of these individuals happens to be a high achiever!

At the other end of the spectrum, there are meetings in which we must appear to be fully engaged even if the meetings are boring and we are only marginally involved. Working on a laptop or other digital devices with tasks unrelated to the meetings is out. I know someone who does breathing exercise whenever he goes to such a meeting. He can still be responsive if needed. Nobody knows what he has been doing when he is not talking. Yet he has charged himself to be more productive for his next task after the meeting. A medical faculty who specializes in hypnosis told me that whenever she is at a faculty meeting, she always puts herself in a self-hypnotized state. I myself often do breathing or isometric exercises.

RULE 5:   Turning spare time into opportunities

Use spare time to engage activities that could enhance our career.

Many people tend to be penny-wise and time-foolish by wasting precious spare time for relatively small monetary saving that is not worth the time.

Like it or not, we technical professionals all face the karma that if we want to advance in our career, we have to work during our spare time. The most notable examples, of course, are the founders of Apple, Microsoft, Yahoo, Google, Facebook, HP, and many other dot.com companies—they all started out humbly, working in their spare time in garages or the equivalent to enhance their careers.

People who spend a substantial amount of spare time on social networking, video games, trips and parties, chores around the house, driving around to find a free parking space, and so on, need to ask themselves: How important is career advancement, how much is their time worth? It is a personal choice and decision.

EXAMPLE: Establishing ourselves as experts

I know many individuals who established their reputation as experts by making presentations at various professional meetings and conferences. Others have achieved the same by writing and publishing articles. More motivated ones have done both. Most of these efforts were done during their spare time. Once they became known as experts, they secured more prominent status and positions in their career. Excellent ROI with their spare time!

EXAMPLE: Expanding our knowledge base

Anyone who has been reasonably successful in his or her professional life is likely to have used some spare time to expand his or her professional knowledge base by self-learning, getting special training, or even earning a more advanced degree. All these are helpful, even necessary, for fast-tracked career growth.

📖 **EXAMPLE: Incubating new ideas and products**

As mentioned at the beginning of this section, the founders of Microsoft, Google, and so on, all started their initial products in their spare time. On a less grand scale, founders of many smaller companies also incubated their initial products by working on them in their spare time. I personally know quite a few people who have been successful financially and/or professionally followed this path.

📖 **EXAMPLE: Focusing on personal enrichment**

There are people who are less interested in moving up management ladders, which is quite stressful. But they can still use their spare time at least for personal enrichment.

RULE 6: Keeping the mind sharp by taking catnaps

Take a short catnap at an appropriate time during the day (only if feasible). Taking a short catnap during the day is probably the best investment of time.

Many researchers have proven that a short catnap during the day can boost alertness, improve memory, enhance performance, and increase creativity. Leonardo da Vinci, Winston Churchill, Albert Einstein, John F. Kennedy, and Ronald Reagan are all known for taking naps in the middle of the day.

Unfortunately, in many environments, taking a catnap still has the stigma of being lazy and could even be a cause for dismissal. In such situations, of course, we have to avoid taking catnaps—unless we can do so during allowed break periods, such as during a lunch break.

Many managers are quite enlightened about the benefits of a catnap. Still, it's best to touch base with our bosses. Based on my limited observations, these managers would not object to a high performer taking catnaps but would frown upon others doing the same.

I personally take about a15-minute catnap in the afternoons. I always had private offices and I regularly put in far more than 40 hours a week, so taking a catnap was never a problem. When I worked as a part-time consultant at two companies during two different summers, I did touch base with the managers about my catnapping habit. They saw no problem with it. But they did make sure I had a private office during the time I worked there.

One last point: the advice of taking a catnap is best suited for professionals who have private offices and who put in more than 40 hours a week to their work. For them, taking a catnap, even during office hours, is really on their own time. Being in a private office, nobody sees them when they take a nap.

◼ **EXAMPLE: FAA decision on air controllers**

A *Federal Aviation Administration* (FAA) policy once forbade air controllers, even on night shifts, to take naps even during their breaks. Apparently, several reported incidents caused by night shift air controllers falling sleep on the job in early 2011, plus criticism about mounting scientific evidence, air controllers on night shifts are now permitted to take naps during their breaks.

◼ **EXAMPLE: "Catnapping" to CEO**

I have a friend, Tim, who "catnapped" his way from being a grunt worker to being president and CEO of a large electronics firm. Rumor has it that when he was still at junior-level positions, his bosses would apologize for interrupting him during a nap. He was an excellent worker yet (or perhaps because) he took catnaps almost daily.

I have known many successful individuals who also take catnaps regularly.

CHAPTER SIX

Career Smart

Principle: "Pond hopping" strategically

Strategy: Find organizations or positions in which our strength gives us the advantage and opportunity

This strategy has us seeking an environment in which we are better qualified than most of our colleagues and there are better chances for us to advance and to be on a fast career track. (Metaphorically, this strategy is akin to being "a big fish in a small pond." Please note that in this sense the "small" is not related the size of the organization. It means a working environment in which our professional skill is more likely to stand out than otherwise.) Many large organizations even specifically have "fast-track" programs for bright new hires, as a way to keep bright staff from leaving and to identify among them promising mid- and senior-level managers.

After we have worked in this environment for a while, there may reach a point where our upward opportunity has become limited (akin to becoming *too* big a fish in the small pond). This is the time to look for a more promising new position elsewhere (hopping to another pond).

Timothy Gaithner, treasury secretary under President Obama's first term, is a good example of someone who, planned or not, has used this strategy well:

- First pond was Kissinger Associates. (A good start!)

- Three years later, pond hopped to be the big fish at the Office of International Affairs at the Treasury Department. (As I will explain later, working at a private company upon graduation for a couple of years before joining the government enables a person to get a much higher salary at a more senior position than otherwise possible.)

Fast-Tracking Your Career: Soft Skills for Engineering and IT Professionals, First Edition.
Wushow "Bill" Chou.
© 2013 The Institute of Electrical and Electronics Engineers, Inc. Published 2013 by John Wiley & Sons, Inc.

- Fast tracked to the position of under secretary in 10 years. (Now, too big a fish!)
- *Pond hopped* to be president of the NY Federal Reserve Bank, a visible and prestigious position.
- Ultimate pond hop to serve as Secretary of US Treasury Department.

He has strategically pond hopped, always to a position in which his strength gave him the advantage and opportunity.

Practicing the following rules, defined later in the chapter, can help us achieve similar deftness in being Career Smart:

▧ Opting to be a big fish in a small pond

▧ Hopping to a more opportune pond at opportune moments

▧ Never polishing a sneaker

▧ Making a good lasting impression by making a good first one.

RULE 1:   Opting to be a big fish in a small pond

Seek to work in an environment where we can stand out. It gives us a better chance to grow in our career.

Someone will always be more or less capable than we are. We just have to play the best game we can with the cards we're dealt. Generally, it's better to work in an environment in which we're the big fish in a small pond, rather than the small fish in a big pond. Being a big fish in a small pond, we're more likely to stand out, we're more likely in a position to broaden our professional and managerial skills, we're more likely to move up the career ladder faster, and we are more likely to be happier. If we grow too big for the small pond, we can then move to a bigger one, where we're likely to jump in at a more preferred position. (The words "small" and "big" here do not refer to organizational size. I use the phrase "big fish in a small pond" to mean someone who shines in his work environment and the phrase "too big a fish" for the situation in which although he shines in his work environment, there is very limited opportunity to move upward.)

For example, if a *typical* new engineering PhD graduate follows tradition and begins to work in a research lab with many accomplished researchers, he would most likely have to work hard just to keep up. If, instead, he takes a contrarian approach and begins to work in a production group, he would likely outshine his colleagues and get a faster promotion. Conventional wisdom thinks that working at a research lab is more prestigious than at a production facility. For *most* people, the reverse is a far better career approach.

(*Caveat*: This comment is for *most* people, but NOT for everyone. Some individuals may simply enjoy, or excel in, doing research.)

🔲 EXAMPLE: Undistinguished career in a big pond

Most talented technical people, especially those with advanced degrees, consider research and development positions to be more prestigious than others. In particular, they prefer R&D labs or research universities. A few people flourish in this environment, but for most people, it's akin to being a small fish in a big pond. They could be happy and proud with what they're doing and have a good career, but it won't be distinguished. Two scenarios highlight the potential hazard of being the small fish in a large pond.

A high-tech startup company had a reputation as a brain trust. The company was able to recruit the best and the brightest. But after 1 year of operation, the company had spent up all its venture capital money, and its income couldn't sustain the entire organization. About half the employees were laid off; several highly intelligent staff (who were akin to relatively small fish in comparison with others) lost their jobs. Their careers were impacted negatively.

Many talented new PhD graduates aspire to work as assistant professors at first-tier research universities, such as MIT or the University of California, Berkeley, where the general rule is that assistant professors can hold their position for a specific number of years, typically seven. If this person isn't promoted within this time frame, his or her contract with the university is automatically terminated. The primary problem here is that the number of positions available for promotion at these first-tier universities is far less than the number of assistant professors who need to be promoted. The net result is that many bright, young people—relatively small fish in comparison with others—get rejected and lose their jobs at the end of a 7-year stint. Had they started at a less prestigious university, they could have been promoted to associate professor.

🔲 EXAMPLE: Shining in a small pond

Some people, whether by design or default, choose to work in less glamorous areas, where they're akin to being the big fish in a small pond and more likely to shine.

Ken, a new computer engineering graduate, was interviewed by two people at a large high-tech company. One was the manager of a software development group and the other was the manager of a testing group. Both offered him a job, but he decided to work in the latter group. At that company, engineers in the development group are generally more skilled than those in the testing group. Ken became an outstanding performer in the testing group. Within 2 years, he received two promotions and big bonuses, which he could not have received had he joined the development group first.

Randy was a newly minted PhD from a top-rated university, starting his career in a high-tech company's development group as a staff member. He was doing well but wasn't distinguished. He then joined another company and worked in its production group, where demand for technical originality was no longer important, and his technical strength afforded an advantage over others. He also has good personal skills, which matter more in production than in R&D. He's now a vice president.

Lou was a computer scientist with a PhD working in a research and development group as a project leader. He then changed his career direction and moved into the IT area as a technical staff member. His technical ability outshone all of his colleagues, including his managers. In less than 5 years, he became a CIO with a US$100+ million annual budget.

Had Ken, Randy, or Lou joined a research lab or a university, it's unlikely any of them would have achieved what they did financially or professionally. They rose to the top because, based on their skills, each chose a "pond" in which he could shine and be the big fish.

RULE 2:   Hopping to a more opportune pond at opportune moments

Having achieved a certain objective at one organization, move to another one for a more promising position.

Depending on our personal strengths, certain environments are more suited for fast career growth than others. However, regardless of where we are, we still might reach the point where it becomes difficult to move up the ladder at the same rate as we did before. This means it's time to move to another environment: (1) we're now too big a fish for our original small pond and need to jump to another pond for better career growth opportunities; and/or (2) we are now qualified for a more coveted or desired position at another pond. A different pond used here may mean a different organization, a different department within the same organization, or even a different project under the same manager.

Figure 6.1 graphically illustrates this concept of pond hopping. This process could be repeated a couple of times during one's career.

The key to success with the "big fish in a small pond" and "pond hopping" strategy as espoused in this chapter depends on how well we can honestly assess our ability ("fish" size) with respect to the potential working environment ("pond" size), as well as our career objectives. That is, *the key to successful execution of this strategy depends substantially on our ability and willingness to choose the right pond.* In 1976, I changed my job from being vice president of the Network Analysis Corporation to being a tenured full professor and the founding director of the Computer Studies Program at North Carolina State University. During his visit in late 1976, UCLA's Leonard Kleinrock asked me bluntly, "How long are you going to stay here? Why don't you come to UCLA?" But I would not have wanted to go to UCLA. I know my limitations and professional objectives. Sure, UCLA is more prestigious than NCSU, and I could have done okay there. But I would have had little chance to shine. I could not have been able to compete with Kleinrock. At NCSU, by working hard, I could make myself a top performer and still have time to do consulting off campus. Of course, I could have picked a less prestigious university than NCSU, but then I would have had less professional clout off campus. NCSU was the right "pond" for my "fish" size, and my strategy proved to be correct. Years later, when applying for a political position as a deputy assistant secretary in the

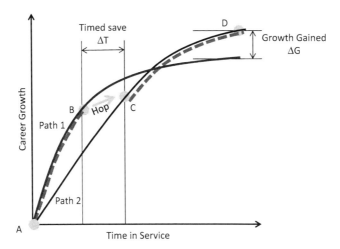

FIGURE 6.1   Concept of pond hopping: Career Path 1 and Career Path 2 represent two typical career growth patterns as a function of time in service. Path 1 has a faster growth initially, but Path 2 has a better career growth in the long run. Point A is akin to being a big fish in a small pond. Point B is akin to too big a fish in the same small pond. A good strategy is to follow Path 1 from A to B as indicated by the dashed line, and then hop to C on Path 2. At a time corresponding to D, the hopping Path ABCD has an advantage over Path 1 by growth gained ΔG, and has an advantage over Path 2 by time saved ΔT.

federal government, I needed recommendation letters from local leaders. Due to my performance at NCSU, the chancellor had been impressed and knew me well. (Had I gone to UCLA, most likely I would not have had such recognition.) I first asked the chancellor for a reference letter, and he wrote a very strong one. Based on that, I asked two congressmen for references. When they read the highly complementary recommendation from the chancellor of the top engineering university in the state, they were comfortable to write similar strong recommendations.

🖥 **EXAMPLE: From industry to government**

A new graduate, regardless of the major, can begin only at a relatively low rank with a relatively low salary in the federal government. (A new graduate with a bachelor's degree begins at Grade 7, Step 1; master's degree at Grade 9, Step 1. In the Washington, DC, area, the salary starts at around $42K and $52K, respectively, as of 2013.) But if he or she goes to work for industry first, he or she can command a much higher starting salary if he or she is in a high-paying field, such as computer engineering. If he or she then goes to work for the government, the hiring manager would have to offer the individual a much higher grade in order to match the salary he or she had been receiving in industry. I know a couple of people who began at General Schedule (GS) Grade 14/15 after working in private industry for a few years. (GS-15 is the highest grade, which is comparable to the colonel rank in the military. GS-14 is comparable to the lieutenant colonel rank.) Without the pond-hopping strategy, they could have not reached these ranks within the same number of years.

Upon receiving his MS degree, Larry worked in private industry as a technical staff member and then as a first-level manager. A couple of years later, he changed jobs to work for the federal government. He was offered a GS-14 position, with a salary matching what he was getting from his job in industry at the time. Had he joined the government fresh out of school, he would have begun at Grade 9, and it would have been highly unlikely for him to reach that level of salary and rank in just a couple of years.

EXAMPLE: From government to industry

For an IT professional, a Chief Information Officer (CIO) is typically the highest position one can expect to achieve at government agencies. Those who have achieved the position of a cabinet agency CIO may find greener pasture in private industry for career growth. Indeed, many large federal agency CIOs retire early from the government and then take a senior position in a private organization. I know one who became a general manager of an IT company and another who became a vice president at a consulting firm.

EXAMPLE: From small company to large company

It's relatively easy to get a fancy title, such as vice president or director, while working for a "small" company, but we also tend to get only "small" responsibilities. Thus, if we want to advance our career, we need to change ponds and move to a larger company or organization. Fortunately, we already have a good title, so we can jump into a good position in the new pond relatively easy.

I've known several vice presidents and presidents of small companies who ended up in large corporations with desirable positions. One president of a small company was given a position as a VP at a very large corporation. One VP of a small company ended up as a division VP at a much larger corporation, another ended up as a director, and a third ended up as a full professor at a prestigious university. Each of these individuals wouldn't have attained these positions within the relatively short time period had they started their careers at these large organizations.

EXAMPLE: From research lab to research university

Working full time at a research lab in a large corporation has two advantages over doing research at universities: corporations are more likely to pursue research related to hot topics. Working full time at a research lab also means we can spend more time researching and therefore producing more results. People who have worked successfully at these research labs have the potential to a senior appointment at universities faster than faculty members who begin their careers at universities.

I personally know of at least five individuals who worked in industry research and development and then moved onto universities. They received a tenured full professor appointment even though they had never been faculty members to that point. Typically, it takes a bright new assistant professor 10 years to make it to a full professor, yet among the five individuals I mentioned earlier, three of them had been out of school for only 7 years. By going into industry and doing full-time research first, they enhanced their chances to become tenured full professors at prestigious universities and shaved 3 years off the time needed to get where they wanted.

EXAMPLE: Strategic multi-pond hopping

I know of several people who by strategically moving from one organization to another more than once were able to map out a fast-track career path for themselves. Sam moved to the federal government as a deputy assistant secretary at one agency, then to another one to become an assistant secretary. From there, he moved to a large corporation as a vice president. He then moved back to the federal government as a deputy secretary.

Willie was working successfully at a prestigious research lab. With this recognition, he was able to get a tenured full professorship at a university in less time than would otherwise be possible. After a couple of years, he "hopped" to another university as a research lab director.

EXAMPLE: Multiple-pond hopping to Senior Executive Service while still in the 30s

The Senior Executive Service (SES) is a federal government civilian career system that is comparable with the military service's general ranks. Few can make it. For those who can, it typically takes most of their career life to achieve that status, and likely occur in their mid-40s or older.

Helen Wood, an IEEE fellow, made it to SES at the age of 38. I interviewed her and tried to discover her secret of success. I found she is good at getting synergy out of multiple rules (as being defined in this book). Here are my takes.

Synergy with rules of "being proactive and farsighted," "showing enthusiasm for challenging assignments," and "hopping to a more opportune pond"
Helen joined the Computer Science Division at the National Bureau of standards (NBS), now called the National Institute of Standards and Technology (NIST), a well-regarded scientific and technical organization among the federal government agencies. NBS/NIST often provides specialized technical assistance to other agencies.

In the early 1970s, the Navy sought help from NBS in developing a real-time control software program for a fuel injection test system. Helen's farsightedness saw that real-time systems had a better growth potential than what she was doing at the time. She jumped the pond (by jumping from one specialty to another) and volunteered for the challenge. She did a wonderful job and impressed her manager and other stakeholders.

Around the same time, the Internet was in its initial developing stage. NBS wanted to take a leadership role in standardizing the communication protocols. This was a totally new area. Nobody had any prior experience. Again, Helen saw the opportunity, jumped ponds, and volunteered to work on the Internet protocols. She proved to have made the right "jump." She continued to move rapidly up through the technical program levels and reached GS-14 only 9 years after graduating from college. (I do not have specific statistics, but it would probably have taken a good performer 15 years or longer to reach GS-14.)

Synergy with rules of "killing two birds with one stone," "successful networking by being networking less," and "hopping to a more opportune pond"
While moving up technical program levels, Helen was presenting her work in outside publications and at speaking engagements ("Killing two birds with one stone"). Through these activities, she accrued good "visibility" and "networking" in the field. (Speaking and publications are two good ways to achieve "successful networking by being networking less.") By properly crafting her presentations, she was able to effectively communicate the value of her work and that of the organization to both experts and those not as familiar with the field. (Being marketing smart!)

But in spite of her "networking" and "visibility," being a technical staff, it was difficult for Helen to move beyond GS-14. She was too big a fish. (In the federal government grade system, it is difficult for a nonmanagerial technical person to be promoted beyond GS-14.) Skilled in communications and marketing, Helen felt that to fast track to GS-15, the next career move for her would be in program and organizational management. So again she "pond jumped," this time into the program and policy office of the agency where she worked to explain their R&D programs to the stakeholders in the government and to Congress. Her skills in communication and marketing matched her perfectly with this position. She started as a program analyst in the agency's Program Development Office, and soon she was promoted to head that office as a GS-15.

But once again, she became too big a fish in the pond. Helen realized that the best path to the Senior Executive Service (SES), that is, becoming a senior executive, in a science and technology agency was to work in the technology management area, not in the policy management area. So she jumped pond again, and accepted a "lateral" move to be the chief of a technical division, where she was soon promoted to the SES at the age of 38. An outstanding achievement!

RULE 3:   Never polishing a sneaker

Move on to another organization/position or give up on seeking career advancement if we cannot advance in our career as we have hoped.

Some of us, at some point, might have found ourselves in a work environment that does not match well with our strength or personal life style. Or, we might have been unfortunate enough to work for a selfish and incompetent manager. When either of these happens, we just have to recognize the fact and stop polishing the sneaker, so to speak. That is, give up trying at the present workplace and find a new job elsewhere. Or, under certain circumstances, simply give up seeking advancement.

This "do not polish a sneaker" principle isn't limited to career advancement. It is equally applicable to many other situations as well, including dealing with the boss, subordinates, clients, and projects. Sometimes we just have to bite the bullet: stop what we have been doing and take a new approach. When Michelangelo was unhappy with his original painting of the Sistine Chapel, instead of "polishing the sneaker," he wiped out the initial one completely before starting over and painting a new one.

EXAMPLE: Manager versus executive assistant

Cathy was an executive assistant for Don, a senior manager. Cathy was very good at working on PCs and was quite sociable. She enjoyed reading tech magazines, and whenever people needed help with their computers, she was there to assist. Whenever Don had a visitor waiting in the reception area, she would readily engage that person in conversation. However, in doing all these things, she was tardy in getting Don's assignments done. She felt that reading PC magazines kept her well-informed, which helped her as she assisted other workers with their computers; she also believed that she should make visitors feel welcome. Don felt that fixing PC problems, especially somebody else's, was tech support's job, not hers. He also felt that entertaining visitors or being well versed in the latest PC news should not have been her priorities—getting his assignments finished promptly was a much higher priority. Rightly or wrongly, Cathy and Don did not see eye to eye and both figured the situation was a lost cause and that they shouldn't even try to "polish this sneaker." While Don was documenting Cathy's activities in order to justify in firing her or getting her transferred, Cathy was looking for another job. Fortunately for both, Cathy accepted a job in another office before Don needed to take any action. They both ended up happy by giving up "polishing the sneaker."

EXAMPLE: Changing jobs

When Randy first received his PhD from a prestigious research university, he started working at an R&D organization, the type of jobs graduates of that university typically desire. In spite of his training, he soon realized that this was not really his cup of tea. In less than 2 years, he decided to stop "polishing the sneaker" and moved to an electronics production company, where original research is not part of the job requirement. He is now flourishing in this new position.

📖 **EXAMPLE:** Wishful thinking and nonchalance

Many of us tend to think things will work out even if evidence indicates otherwise. But if we look around, we can easily identify projects that should be terminated but aren't, people who should have changed their jobs but don't, or businesses that should have closed but haven't. I'm sure some financial wizards saw trouble coming in the economy 2008, but they were either nonchalant or just hoped that things would somehow work out. Such observations should be a constant reminder that sometimes we just have to stop "polishing the sneaker." As President Obama said, "I don't care if you're driving a hybrid or an SUV, if you're driving toward a cliff, you have to change direction."

Paul was VP of Engineering at a high-tech company. Besides the president/CEO, he was the most powerful person in the company. Then the board of directors removed the then president/CEO and appointed a new one. The new president/CEO wanted his own person to be in charge of engineering, so he removed Paul from his position and created a new title for him as VP and Chief Scientist, without any reporting staff. It was very obvious to me that the new CEO was just being kind and did not want to fire Paul outright. He just wanted Paul to find another job and resign from the company. But Paul seemed oblivious to the situation and tried to do a good job in his new position. In less than 2 years, Paul was asked to resign.

RULE 4:   Making a good lasting impression by making a good first impression

Make a good first impression by performing extraordinarily well at the onset of a new position. The first impression is a lasting one.

The underlying assumption of pond hopping as a strategy is that we must be recognized as good performers at every step of the way. To this end, remember a principle that everybody knows: *the first impression is a lasting one.* Whenever we start or move to a new job, we must work especially hard at the beginning to present to our new bosses and colleagues an excellent initial impression.

📖 **EXAMPLE:** First impression, lasting impression

For one job I held, I lived alone during the first 1 month without my family. With nothing to do at home, I spent many evenings at the office. During two of these evenings, my boss had to come back to his office to get something and noticed me still at work both times. He was impressed, and thereafter he told everyone that I was a hard worker, even though I rarely spent my evenings in the office after that first month. (Based on that month's experience, he probably assumed that I worked late at home as well. While I do work late into the night most of days at home, he would have no way of knowing that.)

PART FOUR

Dealing with the Boss

EARNING TRUST AND RECOGNITION

CHAPTER SEVEN

Job-Interview Smart

Principle: Striking a chord with interviewers

Strategy: Prepare well and conscientiously; stay away from any attitude that could be construed as cavalier or arrogant

Anyone who is granted an interview is likely qualified for the job. A successful candidate must showcase that his strength is above and beyond the basic qualifications. The candidate who prepares well and conscientiously is likely to stand out during the interview and stands a better chance to get an offer; a candidate who is cavalier or arrogant during the interview is going to be turned down.

The most important marketing event for career-based individuals is the job interview.

To help identify what it takes to ace an interview, I emailed a friend of mine, Samuel, who gets a job offer after practically every interview, and asked him for his insight. (He is now a division president.) He responded by telling me that his success stems from his "educational background and professional experience," "eagerness to get the job," and "preparedness." (Direct quotes from his email.)

I interpret this recipe for success as follows. "Educational background and professional experience" is the "product" to market, and the better the product, the better chances of getting an interview. "Eagerness to get the job" motivates him to put a conscientious effort into impressing the interviewers, which leads to thorough "preparedness." In short, a successful job interview is a *well-prepared* and *conscientiously* executed marketing task.

I know him well. He has an additional strength he is too modest to mention. He has an amiable persona and *never shows any arrogance* in dealing with people. I am certain that his amiable persona has also contributed to his success.

Fast-Tracking Your Career: Soft Skills for Engineering and IT Professionals, First Edition.
Wushow "Bill" Chou.

The psyche of "eagerness to get the job" is amazing. Every time I was turned down for a position, inevitably, I did not have a strong eagerness to get that particular job. Whenever I did have that eagerness, I got an offer. While I cannot speak for others with certainty, my observation is that I'm not alone in this experience.

Practicing the following rules, defined later in the chapter, can help us achieve deftness in being Job-Interview Smart:

■ Being well prepared by collecting relevant information

■ Putting a positive spin on our qualifications

■ Preparing targeted elevator pitches/speeches

■ Sizing up and resonating with the interviewer

■ Winning interviewers' confidence in us by exhibiting confidence

■ Avoiding gaffes by avoiding overconfidence

■ Being careful of careless comments (refer to Chapter 1, Communications Smart, for a general discussion of this rule)

■ Using plain language (refer to Chapter 1, Communications Smart, for a general discussion of this rule)

■ Inciting enthusiasm with enthusiasm (refer to Chapter 3, Marketing Smart, for a general discussion of showing enthusiasm)

■ Making a convincing presentation by making a well-crafted presentation (refer to Chapter 3, Marketing Smart, for a general discussion of this rule).

In addition, the success of this soft skill can be enhanced by the following soft skills:

• Communications Smart (Chapter 1)

• People Smart (Chapter 2)

• Marketing Smart (Chapter 3).

RULE 1:   Being well prepared by collecting relevant information

Collect and study information relevant to the position, including details about the job, the mission of the organization, and possibly the background of the interviewers.

Interviewers are always impressed with interviewees who have a good grasp of the organization's business and mission and the open position's responsibilities.

In addition, we also want to find information, if possible, on key interviewers' backgrounds and the organization's internal politics. This will help us connect with the interviewers, better respond to their questions, and increase the chance of striking a chord with them.

Nowadays, such information is readily available on the Web, but under some circumstances, we might need to seek help from our network of professional contacts.

RULE 2: Putting a positive spin on our qualifications

Give ourselves a positive spin by playing up our strengths and playing down our weaknesses, especially with respect to the position for which we are applying.

Information collected about the organization and our interviewers can be helpful here.

The fact that we're given an interview indicates that we have the right educational background and professional experience—but so do all the other interviewees. We can use positive spin to impress upon our interviewers that we're more valuable than the application form shows.

For example, if we're long in technology and short in management experience, we should make a case as to why it's crucial that the position be strong in technology. If we have more management experience, we should showcase the success of projects under our management to stress how effective supervision trumps being a technology expert. If we're experienced in both, we need to emphasize the value of having an experienced manager with technical expertise.

Once I needed to recruit a couple of techies to tackle a number of technical problems. My staff formed a recruitment committee, and after a round of interviews, submitted to me a list of top 10 candidates to choose from. It turned out that nine out of the 10 were not techies, but they all had experience in successfully managing large technical projects. They were apparently good at putting a positive spin on their qualifications by convincing the recruitment team that to tackle technical problems, technical management experience is more important than actual technical skills.

RULE 3: Preparing targeted elevator pitches/speeches

Prepare and craft a set of "elevator speeches" specific to anticipated questions, with the objective of delivering a positive spin and enabling resonance.

We may or may not be asked to give a formal presentation, but an interview always consists of a set of Q&As. Each "A" is akin to a mini presentation, or an "elevator speech." These questions are mostly from a boilerplate. (Various versions of such questions are readily available on the Web with a simple Google search.)

In most situations, these questions fall into two categories. The first is mainly about ourselves: our strength, weakness, perception of our current position, perception of the job we are applying, the reason why we are seeking a new job, our career goals, and so on.

The second set relates to the organization and the position that is being filled. For example, if we are interviewed for a technical position at Google, we should not be surprised if we are asked about questions on data search. That was exactly what happened to Marisa Mayer. During her interview, she was asked how she would handle the large database associated with spell checking. (Marisa Mayer began her career at Google as an engineer and quickly moved her way up to senior executive. In 2012, she was recruited by Yahoo as its CEO.)

Prior to an interview, we need to anticipate potential questions and prepare a set of elevator pitches/speeches that will, hopefully, resonate and deliver a positive spin. One useful strategy is to prepare and carry with us a set of relevant charts with graphs and talking points before an interview. The charts may then be used to facilitate our presentations.

RULE 4:   Sizing up and resonating with the interviewer

Tailor our statement to the interviewer's interest and expectation with the objective of inciting resonance with him/her.

Some resonating remarks have to ad libbed, but many can be prepared prior to the interview, particularly if we have collected relevant information about the organization and the open position. This information can help us prepare our set of elevator speeches, including putting a positive spin on our qualifications, and sharpening our vision for how to carry out our prospective responsibility successfully, all of which will resonate with the interviewer.

An IT manager working in industry was applying for an IT managerial position in a federal agency. He knew that the government outsourced a substantial amount of its IT work, and he assumed there were grievances with the vendors. So, he prepared answers to questions on this topic (*elevator speeches*). Sure enough, a question about dealing with outsourcing vendors came up. In his response, he put a *positive spin* on how he had successful dealt with such problems. His response *resonated* well with every member on the recruitment committee.

There are situations in which the same comment can resonate with one interviewer and upset another. Sometimes we may even face them at the same time. Such situations do not happen often, but when they do, we need to make our positions and viewpoints more flexible. Resonance may have to take a back seat to avoiding irritation. For example, if we apply for a CIO position at a university, we'll likely face some faculty who are strong advocates for e-learning (computer-based instruction), as well as others who are strong adversaries. If we express our support for one of the two disparate strategies, we would surely offend one group. To avoid this unpleasant consequence, we may contend that e-learning can play an important role for some subjects, but not all. Responses like that may not resonate well, but won't offend anyone either. Not offending is more important in situations like this. If we apply for a CIO position in industry, we may encounter different views on cloud computing. Again, we may need to make our comments on the topic gingerly.

RULE 5: Winning interviewers' confidence in us by exhibiting confidence

Present ourselves with confidence, and we will gain confidence from interviewers.

I know of an otherwise highly qualified person who did not get a senior management job because he did not show enough enthusiasm during the interview. I had recommended him to a CEO of a large international corporation for a CIO position. The CEO explained to me why he did not make an offer to the candidate: lack of "dynamism," in other words, not showing enough *outward excitement, confidence, and enthusiasm.*

RULE 6: Avoiding gaffes by avoiding overconfidence

Be vigilant not to step from confidence into overconfidence.

There's a thin line between being confident and being overconfident. A confident person leaves a good impression; an overconfident person tends to be arrogant, opinionated, and prone to making gratuitous insensitive remarks that trigger negative feelings. This is a particularly important precaution for successful people. Somehow, successful people are especially susceptible to being arrogant.

During a job interview, we do want to show our strengths and know-how, but we should be especially careful not to point out an interviewer's weakness or to be too opinionated. We want to accommodate an interviewer's viewpoint, not win an argument. We need to be aware that showing off could backfire and turn some people off.

STORIES OF FAILED INTERVIEWS

The failures described in the following stories can help us avoid similar mistakes in our own job interviews. Lessons learned from failures are often far more valuable than lessons learned from successes.

EXAMPLE: Lacking preparation

I was once offered an interview for the position of Assistant Secretary for Technology Policy at the Department of Commerce. In hindsight, I was woefully unprepared for the interview. (The only excuse I had was that I was given a short notice for the interview and at the same time I was busy on another project.)

The person who would interview me was the Deputy Secretary (DS). I did not know much about him, but I did know someone who worked very closely with him. I should have called my friend to inquire about the DS's background and his current policy interests. But I did not. I should have searched the Web for the DS's speaking topics. But I did not. Anyone who was being interviewed for that position should have had some knowledge and opinions about hot technical issues at the time. I should have made preparations for these issues. But I did not. At that time, a hot topic for both IT professionals and other policy makers was the so-called Y2K bug. I should have thought of that and prepared a couple of good elevator speeches on the topic. But I did not. During the interview, how to handle the Year 2K problem was the first question I was asked. I did not answer well, particularly from the perspective of making a policy. It was a poor start. From that point on, the whole interview proceeded more like fulfilling a boring obligation than sizing up a potential candidate.

   (*Lessons learned:* **Always** *prepare well for an interview.*)

⚑ EXAMPLE: Unguarded comments

A candidate for a college deanship was having dinner with the college's department chairs. He made an unguarded comment that department chairs should not give up conducting research. But several of the chairs present did very little research, and of course, this comment offended all of them. These department chairs voted against him, and he didn't get an offer.

   (*Lessons learned: Part of the problem was that he did not prepare well for the interview. If he did, he would have noticed that at that university, most department chairs did not do research, and he might have been more guarded in making such remarks.*)

⚑ EXAMPLE: Expressing disagreement with interviewers

I was once asked for a second visit for the position of vice president at a university. Several people I talked to during that return trip implied that this job was mine to accept or reject. As the day went on, my self-confidence continued to grow, and then came the dinner. Present were the university president, his wife, and senior university officials. During the dinner, I was told that they would expect me to be loyal to the president if I were offered the position. Without any hesitation, I responded by saying my loyalty would be with the university. It was immediately clear that they did not like my response. One week later, they called me and turned me down with a very flimsy excuse—due to certain internal constraints, they could not pay me the salary that they thought my qualifications deserved. Since I did not make any definitive salary request, this could not be the real reason, which, I believe, was my arrogance displayed during that dinner.

   (*Lessons learned: Obviously, I should have been more agreeable. If I did not like "being loyal to the president" instead of the university, I had the option of declining the offer when given.*)

▣ EXAMPLE: Unnecessary arrogance: I

Dan is a very well-known achiever in the computer field. In addition to being a member of the National Academy of Engineering, he has many patents, developed widely used software systems, invented award-winning hardware, and published many papers. When a research university tried to recruit some high achievers to its faculty by creating chaired university professorships, Dan applied. Obviously, he was very qualified, and the computer science faculty were excited to have such an individual in their midst. But he could be arrogant from other people's perspective. While he could have presented a multipage-long resume listing all his achievements, he sent in only a short page. He was asked to send a more detailed resume, but he refused, saying that he did not have one. From his viewpoint, he had achieved so much, that a detailed listing of all his publications, awards, inventions, and so on, were irrelevant and therefore he never bothered. This irritated the recruitment committee, particularly the committee chair. The committee was ready to reject Dan's application.

At the last minute, the Department of Computer Science sent a top faculty member to talk to the committee. This person explained Dan's achievements patiently to the committee and convinced them that it would be to the university's advantage to offer Dan the position. Finally, the committee consented to offer Dan the appointment. This is the only exception I know of in which a seemingly arrogant applicant succeeded.

I am sure Dan himself does not think he is in any way arrogant. He probably thinks he is quite unpretentious. I know many high achievers act somewhat arrogantly in others' eyes, but they do not think so themselves.

(Lessons learned: While Dan earned his right to be seemingly arrogant, it would not take much for him to produce a more detailed resume. There was no reason to risk being rejected. For most of us, we need to be particularly aware not to be unnecessarily arrogant, particularly from other people's perspective.)

▣ EXAMPLE: Unnecessary arrogance: II

While I knew Dan's case was an exception, I nevertheless took a similar attitude.

When I applied for the position of the presidency of a small university, the recruiting committee responded promptly, writing to me and calling me twice. But then they asked me to do something I thought was ridiculous: they wanted me to prove that I indeed graduated and received the degrees, GPAs, and class rankings as I had claimed. They asked me to have these schools send them transcripts to verify my degrees and my claims. I adopted Dan's attitude and refused. From my perspective, unnecessarily arrogant, what I had achieved in my career rendered irrelevant my school records. But I missed a key point: that university is far less sophisticated than a research university. The recruitment committee probably did not have the ability and/ or confidence to evaluate a resume's veracity. Apparently, they depended heavily on official documents to handle that task. While I thought their request was ridiculous, they probably thought I had something to hide. I was turned down for refusing to comply with their requests.

(*Lessons learned: There was no need to be unnecessarily arrogant. I could have asked my alma mater to send transcripts to them.*)

EXAMPLE: Unnecessary arrogance: III

Strictly speaking, the next two stories are not failed interviews but are about lost bets from a "double or nothing" strategy.

Many years back, the directorship of the National Institute of Standards and Technology (NIST) was open. Through some contacts, the White House (WH) personnel office sent my resume to the Under Secretary of Technology Policy at the Commerce Department (NIST is under the supervision of the Under Secretary). The Under Secretary contacted me and offered an interview. Apparently, she was impressed with me. Because of that, the WH personnel office became very enthusiastic about my candidacy. Within a couple of months, the situation changed. A new Secretary of Commerce was appointed, and the Under Secretary decided to return to private industry. The new Secretary had unofficially promised the position to someone else, so the WH personnel office asked me whether I would be interested in the deputy director position. The sign was clear. If I were willing to consider that position, it was mine, and everybody would be happy. If I were to insist on applying for the director position, there would be more paperwork on all sides, and my chances of getting the position would not be zero, but would be small. Furthermore, there was an indication that the other fellow would not finish his term, and I might have a better chance for another try. I was not particularly interested in the deputy director position. I also arrogantly believed I was more qualified for the directorship. So I took the "double or nothing" strategy. As expected, I did not get the directorship position.

In another situation, the chair of a dean search committee at a university contacted me through a mutual friend and asked me to interview before the committee officially began its search. I did, and the interview went very well. The committee told me the position was mine to accept or reject, but they had a small request: if I wanted to accept the position, I should do so in a few days to save them the trouble of a formal search process. One thing made me hesitate to accept the offer: the presidency position was also vacant, and more than one person told me that I was more qualified than any of the three finalists. Unfortunately, the deadline for applying for that position had already passed. The only chance I might have for being considered for the presidency position was to write to the chair of that recruitment committee and plead for a deadline extension. My arrogance again took me to the path of "double or nothing." I declined the deanship offer and sent my plea to the chair of the president recruitment committee. As expected, the chair responded that he could not make an exception.

(*Lessons learned: Due to personal reason at the time, I did not particularly care about the deputy director's or the dean's position. Had I cared, I should not have taken the "double or nothing" strategy.*)

EXAMPLE: Showing off

A CIO candidate was visiting an IT shop's security technology center. He immediately noticed its inadequacies. Without mincing words, he pointed out all the weakness and said what should have been done. The staff became embarrassed and defensive.

(*Lessons learned: The candidate could have softened his critiques and suggestions; instead, his showing off led him to make insensitive remarks.*)

EXAMPLE: Gratuitous comments

I was interviewed for a dean's position at a university located near a scenic vacation spot. During a meeting with a roomful of faculty, I happened to mention I turned down an offer from another university also located near a vacation spot. I then made what I thought was a jovial remark: "I did not waste my time. At least I had a free vacation." I noticed instantly from their facial reactions that several faculty were not amused by my joke and took it quite negatively. The damage from this gratuitous, unguarded remark was done.

(*Lessons learned: When we interface with a large number group, we should be aware of the fact that different people have different senses of humor. What we conceive as a joke could be a turnoff to others.*)

EXAMPLE: Broadness versus depth

An ideal senior technical manager is expected to have some knowledge of a broad spectrum of relevant topics, while a senior technical staff is expected to have an in-depth knowledge of his area of specialty.

Jon was applying for a faculty position at a research university where a faculty member was expected to be an authority in his field of specialty and to publish papers of original research and obtain research grants. During the interview, Jon claimed to be equally good in multiple areas. This claim did him in. The recruitment committee unanimously rejected him, reasoning that it's humanly impossible to be authoritative with in-depth knowledge in multiple technical areas. They felt Jon overly exaggerated his strengths and was not credible.

(*Lessons learned: We need to be careful not to oversell ourselves.*)

📓 EXAMPLE: Salary negotiation

A professor was starting a high-tech start-up company. He offered positions in the company to four of his students: Abe, Bob, Charlie, and David. Abe, Bob, and Charlie all immediately accepted their offers. David, on the other hand, asked for a higher salary. The professor could not do that. If he did, he would have to adjust the salary for Abe, Bob, and Charlie as well. As a result, he did not extend an offer to David.

Jeff, who had been working at a government agency, was offered a more senior position at a different agency. The government has specific rules in adjusting salaries when one moves from one position to another. In Jeff's situation, however, the interpretation of rules was not clear-cut, so Jeff took a hard position that he would only accept the position if the interpretation was to his liking. Since the process would involve human resources staff at two agencies, it would take a while to get it resolved. Because there was no assurance the final interpretation would be to Jeff's liking, the hiring manger rescinded his offer to Jeff.

(*Lessons learned: While we all want to get as a high salary as possible, the hiring manager always has constraints on what he can offer. We have to be careful not to force the hiring manager, as well as ourselves, into a corner.*)

A SUCCESSFUL INTERVIEW STORY

At the beginning of this chapter, I pointed out that Samuel has had a very high success rate in getting an offer. In his own words, he credits his success to his "eagerness to get the job."

For various reasons, I had a strong "eagerness to get the job" as Deputy Assistant Secretary for Information Systems (DASIS) at the US Department of Treasury. (The DASIS position was similar to that of a CIO.) At that time, that position was likely to be the most challenging one among all CIOs or equivalents. It oversaw a very large IT budget. (It was likely to be larger than any nonmilitary organization.) It had both substantial oversight and operational responsibilities. (A typical CIO or equivalent has only one or other.) It was the only DASIS that was a political appointee. (At other government agencies, the DASIS position or its equivalent was a career one.)

The odds were heavily stacked against me. For a political appointment position, I had little political connections. Past incumbents had always been nontechnical, and I was technical. People at this level all have excellent verbal skills, and I was a foreign immigrant from a non-English-speaking country.

In spite of these hurdles, I had an interview and an offer. This surprised many people. I have been asked by countless acquaintances how I got the appointment. All I could say was, "Where there's a will there's a way."

How to get an interview is not the theme of this section, and I will not belabor it. But suffice to say that I did pull lots of strings to get my resume on to the desk of the hiring manager (Assistant Secretary for Management, ASM), and I designed an impressive resume to impress him. It must have worked because it led him to contact me for an interview.

Collecting relevant information through "networking"

Prior to my scheduled interview, I spent days collecting relevant information on the Web and from my professional contacts. I learned the following:

- Organization (Deptartment of Treasury): 14 bureaus, ranging from the IRS to Secret Service.
- Responsibility of the position (DASIS):
  1. Overseeing 14 bureaus' IT resources (14 CIOs, 11,000 staff, $2.0B annual budget)
  2. Managing a $285M IT budget, including a 4000-node communications system serving 140,000 users
- Interviewer (ASM): a Harvard-educated lawyer with an MBA, very active in politics.

I also collected, through my professional friends, documents relating to various IT systems at these 14 bureaus.

All this information helped me tremendously in preparing for the interview.

Preparing targeted elevator speeches

Although an interview isn't really a presentation, I viewed each answer as a micro one. I anticipated the type of standard questions I could be asked. Based on the information I collected earlier, I prepared three diagrams (for more details of the three diagrams, please see the example "Three diagrams impress a job interviewer" in Chapter 1, Communications Smart).

I also prepared three charts:

- One bullet chart summarizing key IT issues facing Treasury
- One bullet chart summarizing evolving technology
- One graphic chart summarizing DASIS's responsibilities and it relationship with 14 bureau CIOs

Depending on the questions, I prepared myself to answer them by using one or more of these diagrams and charts. These graphics helped provide simple, *understandable* explanations to even complicated technical problems in a very limited time frame. It was obvious that ASM was impressed.

Positive spin

I explained to ASM why the person holding the position as DASIS should have strong experience in both management and technology. This of course was a spin on my strength.

I was aware I did have an obvious weakness. My English skills were not as polished as other senior officials at a similar level. To put a positive spin on this,

I brought a list of professional meetings where I was a frequent paid speaker, proving that I had more success in conveying complicated concepts to the audience than my "English as a first language" counterparts. I also brought examples of my writing to prove that I was very good at conveying ideas in writing as well. I successfully made the point that the ability to convey is more important than polished writings and speeches. Again, this was a spin on my strengths and weaknesses.

Resonance with a vision

Knowing ASM's background, I surmised that he would want to leave a legacy and that he would want to have a good understanding of IT systems within the Treasury Department. To address the legacy objective, I presented myself not just as a caretaker for the IT systems but as a strategist and visionary with a plan for advancing IT systems at Treasury to enhance the department's mission. The diagrams and charts helped me explain this vision.

To make ASM understand the IT systems, I succinctly explained the basic concept of the Treasury's IT system using nontechnical terms. Again, the diagrams and charts helped. This really resonated well with ASM. Until then, he didn't have the faintest idea what was going on with the IT systems, despite weekly reports from senior IT staff. ASM was particularly impressed with my ability to understand how the various systems interacted in spite of not being a Treasury employee. (My effort in collecting relevant information prior to the interview paid off.)

Showing confidence

My experience as a frequent public speaker and presenter gave me an extra edge. I was used to making thoughtful comments, and throughout the interview process, I exhibited confidence and measured enthusiasm about what I could achieve. I was mindful that I needed to leave an impression that I am a doer, visionary, and team player.

CHAPTER EIGHT

Boss Smart

Principle: Making the boss look good

Strategy: Allow our boss to take credit and benefit from what we have done; promote his/her policy, and shield his/her weaknesses

Cynically speaking, our main job responsibility is to please our boss and to make him/her look good. This entails we perform well in a way that our boss can claim credit for and to be loyal in a way that our boss's policies promoted and his weaknesses shielded. Sometime, we may even need to shoulder blame for our boss's fault.

Gen. Stanley McChrystal and Gen. David Petraeus demonstrate well the opposite ends of this doctrine.

Gen. Stanley McChrystal: In 2010, when Gen. Stanley McChrystal was in command of military operations in Afghanistan, he made disparaging remarks about the Obama Administration and, consequently, *made his bosess look "bad."* He was immediately relieved of his command and forced to retire. (Euphemism: he resigned the post of his own volition.) He paid a big price.

Gen. David Petraeus: At the time Gen. McChrystal resigned, the war in Afghanistan was at a critical point. The whole situation put the Obama Administration in an awkward situation. Gen. David Petraeus seemed to be the only person who had the ability and prestige to take over the command, but he was Commander of Central Command, a higher position than the one in Afghanistan. As a good loyal soldier, he gladly took over the lower-ranked position. This solved the Administration's headache and changed the Administration from looking bad to looking good.

Fast-Tracking Your Career: Soft Skills for Engineering and IT Professionals, First Edition.
Wushow "Bill" Chou.
© 2013 The Institute of Electrical and Electronics Engineers, Inc. Published 2013 by John Wiley & Sons, Inc.

He did an excellent job in carrying out the Administration's policy in Afghanistan, as well as openly promoted and supported the Administration's policy. That is, he continued to *make his boss look "good."* Fewer than 10 months after he assumed the Afghanistan command, he was nominated to be Director of the CIA. A fast-track promotion! He was *amply rewarded for making his boss look good.*

Practicing the following rules, defined later in the chapter, can help us achieve deftness in being Boss Smart:

■ Winning trust by showing loyalty

■ Gaining gratitude by sharing credit and taking blame

■ Being astute by watching for nuances

■ Being proactive and farsighted

■ Showing enthusiasm for challenging assignments

■ Inciting enthusiasm with enthusiasm (Refer to Chapter 3, Marketing Smart, for a general discussion of showing enthusiasm)

■ Being careful of careless comments (Refer to Chapter 1, Communications Smart, for a general discussion of this rule)

■ Using plain language (Refer to Chapter 1, Communications Smart, for a general discussion of this rule)

In addition, the success of Boss Smart can be enhanced by the following soft skills:

- Communications Smart (Chapter 1)
- People Smart (Chapter 2)
- Motivating Smart (Chapter 9)
- Delegating Smart (Chapter 10)

RULE 1: Winning trust by showing loyalty

Use actions and words to exhibit our loyalty to our bosses.

For senior-level positions, loyalty is often the most critical criterion for getting an appointment and staying on the job.

Our culture is such that we're expected to be loyal to our bosses. The more senior the position, the deeper loyalty is expected. In Vice President Joe Biden's acceptance speech as the Democrats' VP nominee in September 2012, he only spent 10 words to highlight his own strength: "President Obama knows the depth of my loyalty to him."

Among all the soft skills, loyalty is the trickiest. In most cases, it doesn't cause problems and even leads to handsome rewards, but there's a caveat: some bosses expect us to protect their interests ahead of the organization or institute for which we both work. At best, they expect us to drink the Kool-Aid under any circumstances; at worst, they expect us to be the fall guy if things go wrong. *We must recognize, and not cross, the fine line beyond which we compromise our own integrity* or, worse, end up violating ethical, moral, and/or legal constraints. If we are forced to cross this fine line, we may have no choice but resign. In 1982, that was exactly what Chuck Hagel did. He resigned as Deputy Director of Veterans Administration because he disagreed with the Director's policy. (As of 2013, Chuck Hagel is the Secretary of Defense. He was a US senator from Nebraska.)

EXAMPLE: Fired for being disloyal

The president of a high-tech company had regular meetings with his company's VP of marketing. Whenever a difference of opinion arose, the VP always expressed agreement with the president congenially, but after the meeting, he usually ignored the president's instructions, and did things his own way. The company's president eventually caught on—considering the VP to be both disingenuous and untrustworthy, he summarily fired him.

EXAMPLE: Fired for upholding integrity

In 2003, US Army Chief of Staff General Eric Shinseki was called upon by Congress to testify about the war strategy in Iraq. His viewpoint was quite different from the policy strongly espoused by the then-president and secretary of defense. When questioned, he could either hide his own opinion and compromise his integrity or testify honestly and offend his bosses. He chose to be honest to himself. A few months later, he was forced into retirement. (Although the Bush Administration probably didn't like him, the Obama Administration apparently holds him in high regard for his integrity—he was appointed as Secretary of Veteran Affairs.)

EXAMPLE: Derided for lacking both integrity and loyalty

In 2008, Scott McClellan, White House press secretary from 2003 to 2006, wrote a book entitled *What Happened: Inside the Bush White House*, in which he criticized President George W. Bush, the White House, and the Washington culture during Bush's tenure. As expected, he received scathing criticisms from conservatives, but he didn't earn compliments from liberals either. Although many on the left were delighted in the book's contents, they couldn't praise someone who was judged as lacking both integrity (for staying on as White House press secretary) and loyalty (for writing the book and openly stating his criticisms).

EXAMPLE: Quandary during job interviews

Occasionally, a hiring manager may ask a candidate whether he would be willing to pledge loyalty to him if offered a position. For some people, personal ego makes it difficult for them to pledge loyalty to a stranger; others have no such difficulties. It's purely a personal choice, but it also depends on the culture of the organization for which we work. In one position I held, some of my direct reports expressed their loyalty to me during my first one-on-one meetings with them; it might seem strange, but it was that organization's culture.

RULE 2: Gaining gratitude by sharing credit and taking blame

Share credit with our boss for projects we have successfully completed and take the blame when what the boss has contributed has failed.

If we share credit with our boss, we lose nothing and gain our boss's appreciation. If we shoulder the blame for some failure to which our boss has contributed, we usually have little to lose and gain our boss's gratitude. (As always, there's a caveat: watch out if our boss expects us to be the fall guy for him. Obviously, it could bring about serious consequences. We have to decide whether we want to be a "John Dean" or a "Scooter Libby.")

EXAMPLE: Consequence of sharing versus not sharing credit

Rob and Joe were highly competent technical professionals, working within the same division at a company. They had a completely different attitude about sharing credit. When Rob made a technical presentation to the division VP, he went to the VP directly without going through his manager. He wanted to make sure the VP knew that all the work was his effort alone; he didn't want to share any credit with his manager. When Joe made a technical presentation to the VP, he always asked his first- and second-level managers to go with him. By doing so, he implied that he did his work under their supervision and guidance and shared the credit with them. Guess who got promoted much more quickly? Rob ended up taking an early retirement in frustration, feeling underappreciated, and Joe is now a senior executive.

EXAMPLE: Taking the blame

The VP of Telecommunications asked a software manager to develop a simple program for an application, but they had differing opinions on which algorithm should be implemented. The manager yielded to the VP's approach, but the algorithm favored by the VP didn't work out well, and the software was a flop. In reporting the progress at a staff meeting, the manager took all the blame himself with no mention at all about who recommended the algorithm. Although the VP never said anything, we can assume he was appreciative and the manager would be duly rewarded.

RULE 3: Being astute by watching for nuances

Be astute in detecting nuances in our boss's expressions and "between the lines." From nuances, we can determine his/her real likes and dislikes and what he/she really wants done or not done.

Managers, particularly senior ones, might not always verbalize what they have in mind. They might not vocalize their displeasure with what we've done because they don't want to embarrass us or they feel that expressing their displeasure wouldn't ultimately help the matter. Or, they might not verbalize what they have in mind because they haven't finalized their decision yet. Regardless of the reason, this is our opportunity to impress them by proactively doing what they have in mind. This is also our opportunity to influence their mind before they verbally finalize their thoughts and decisions.

EXAMPLE: Consequence of being insensitive to nuances

It puzzles me when many otherwise smart people can't read the writing on the wall that they're going to be in big trouble. If they could, they would either immediately improve their performance to avoid being fired or demoted, or they would look for and land a new job elsewhere.

Conan was a VP in charge of a product line at a large electronics company. During the senior VP's staff meetings, the SVP mentioned many times over the course of 2 years that the revenue in Conan's group was far lower than in any other groups. You would think Conan would either adjust his product line to improve revenue or look for a job elsewhere, but he did neither. When the SVP reorganized the groups under her charge, Conan became a VP without portfolio—nobody reported to him, not even a secretary. You would think Conan would find a job elsewhere immediately after this change occurred, but he didn't. After a year, he was shown the door.

Dave was a technical staff member at a municipal agency. During two consecutive performance reviews, Dave's manager made critical comments about Dave's performance and behavior. It was obvious that the manager was documenting a paper trail to fire Dave, yet Dave seemed to be oblivious to the nuances. When Dave's manager finally asked him to resign voluntarily, Dave was surprised.

EXAMPLE: Heeding nuances: I

A day before an interdivision committee meeting, the SVP who chaired the committee asked a senior staff member, Ed, to make a presentation for it. One hour before the meeting, Ed showed an outline to the SVP. The SVP approved the content without much comment. But Ed detected that the SVP wasn't exactly happy and knew it couldn't be the content: if it were, the SVP would have made some suggestions. It had to be something the SVP didn't think there would be enough time to fix. Ed figured the SVP wasn't happy with the presentation's format—he wanted a dog-and-pony slide show. Luckily for Ed, he was good at PowerPoint, and he had several templates for different occasions. He returned to his office, told his secretary to hold all his calls, and worked feverishly on a slide presentation. The SVP couldn't believe that Ed was able to convert an outline sheet into a dog-and-pony slide show in an hour, but it goes without saying that he was utterly delighted.

◨ EXAMPLE: Detecting nuances early to influence the final decision

When the US Congress passed an act to establish a CIO position at cabinet agencies in 1996, it didn't provide clear-cut instructions on how to implement this change. The person who had the most influence for actual implementation in most agencies would be the Assistant Secretary for Management (ASM). This person would make the recommendation to the Secretary about how to implement the CIO position. There were three main choices:

1. The CIO could be at the same level as the ASM and hold an assistant secretary position as assistant secretary and chief information officer (AS/CIO).
2. The CIO title could be assumed by the ASM to become ASM/CIO with a deputy CIO serving as the actual person responsible for IT.
3. The CIO title could be held at a deputy assistant secretary level as Deputy Assistant Secretary and CIO or DAS/CIO with the DAS/CIO reporting to the ASM.

Clearly, many individuals who held the ASM position wanted to assume the CIO title as well. One such individual was thinking in this direction but hadn't made a final decision. He asked the DAS for Information Systems, Gene, who had the equivalent responsibility of a CIO, to write a report to evaluate these three options. Gene knew his boss would absolutely not accept option 1, so his report listed many of its disadvantages to give the ASM the ammunition for recommending to the Secretary to reject this option. Although Gene knew his boss would prefer option 2, Gene absolutely hated it, because Gene would then be Deputy CIO instead of CIO. Gene figured option 3 was an acceptable compromise, so his report emphasized option 3's advantages over option 2. Gene's strategy worked: his boss adopted option 3 as the final recommendation, and Gene was appointed as the DAS/CIO.

Gene was able to read his boss's mind and worked out a solution acceptable to both before his boss made the final decision.

◨ EXAMPLE: Heeding nuances: II

At Corporation N, each division had its own CIO and data centers. The CIO at the headquarters, who coordinated all the IT activities at Corporation N, felt that both cost and reliability could be enhanced if some of the data centers were consolidated. To do so, he needed buy-ins from division presidents. Edward, one of the division presidents, asked his CIO, Frank, to present him with a pro and con analysis.

Frank in turn asked a senior manager, George, to make an analysis for him. Frank did not like to lose direct control of data centers currently operating in the division. He could not openly express this somewhat self-serving viewpoint. Reading between the line nuances, George figured out what his boss wanted to hear. He wrote a report that emphasized the disadvantages if the division was to lose direct control of the data centers, and downplayed any purported advantages of data center consolidation. George's report really pleased his boss!

RULE 4: Being proactive and farsighted

Actively look for ways to improve performance, expand services and functionalities, and increase return on investments.

Managers value subordinates who are proactive and farsighted. They can count on a competent, proactive, and farsighted staff for performance that would live up to or exceed expectation without much supervision. The crucial point here is that *the action and the vision as a result of being "proactive and farsighted" must resonate well with the boss.*

EXAMPLE: From first-level manager to third in 2 years

Almost 20 years ago, Henry was a first-level manager at a large computer firm. At that time, high-speed communication as we know it today was still in its nascent stage. But Henry realized its importance and pushed the idea to his bosses. He was able to market his idea well. Consequently, he was assigned as a product manager to start up a new group related to broadband communications products. As broadband communications expanded, so did Henry's responsibility. Within 2 years, his group soon expanded to three levels, and he was appointed as the director of the group, a third-level manager.

EXAMPLE: Farsightedness not resonating with the boss

To succeed, farsightedness has to resonate well with the boss. In the above example, Henry did that, but I know of a case where farsightedness did not work. Jorge was a senior engineer working on a new product for an electronics company. He expressed his interest in being a manager to the VP in charge of the product. The VP asked Jorge to propose some ideas that could enhance the functionality of the product that was being developed, but Jorge could not propose any ideas that resonated well with the VP. He did not become a manager, in spite of his efforts.

RULE 5: Showing enthusiasm for challenging assignments

Be willing to accept any challenging assignment from our boss and perform it well. This is the opportunity to get recognition and promotion.

When our boss assigns a challenging task to us, we should always be ready to accept the assignment whether we know the skills needed to perform it or not—as long as the required skills fall within the radius of our expertise. If we are not familiar with the required skills, we can learn them, broaden our expertise, and appear to be a positive and willing contributor. We should consider this as an opportunity for increasing our chance to advance our career.

EXAMPLE: Accepting challenging assignments to get a fast promotion

Around 1970, Jon, a new graduate with a PhD, joined a consulting firm as a senior technical staff member. At the time, large-scale data communication networks were developed rapidly. Yet no one seemed to know how to analyze and configure such networks cost-effectively. This consulting firm decided to pursue contracts to tackle exactly such problems, even though nobody in the company at the time knew how such networks worked. Shortly after Jon's arrival, he was sent to see a client for a potential contract to analyze and configure the client's data network. Just like everyone else in the company, Jon didn't know anything about communication protocols or other characteristics of such networks, but he accepted the challenging assignment without hesitation. He read documents about these types of networks prior to his visit with the potential client. He then learned more details from the client. When Jon delivered his report and recommendations, the client was utterly impressed and was convinced Jon was an expert on the problem. In fact, Jon had not been an expert at the inception of the project, but he was at its conclusion.

Because of Jon's impressive performance, the company continued to assign him to challenging projects. Jon was always willing to accept the challenges and work on different types of data networks that he had no prior background with. In less than 3 years, he became well known in the field. During this period, he was promoted from a senior technical staff member to a director and finally to vice president.

PART FIVE

Dealing with Staff

INSPIRING LOYALTY AND PRODUCTIVITY

CHAPTER NINE

Motivating Smart

Principle: Caring for staff

Strategy: Keep in mind the best interests of staff's career and act accordingly

A manager, for the most part, can only be as productive and creative as his staff. A motivated staff tends to be more productive and creative. A sure way to motivate staff is by genuinely caring about them and building a good rapport with them.

Winston was regarded in his company as the senior manager with the most productive staff. His colleagues attributed this to Winston's ability to recruit a good team. This was only partially true. The main reason, I believe, was Winston's *caring* for his staff, which motivated them to be productive. The following statements made by his former staff reflected the *rapport* he and his staff shared. Within a year or so after Winston left his post, several staff made nostalgia remarks about him: "Since Winston left, it is no longer fun working here. The new guy is a stuck-up. I am taking an earlier retirement," "Winston is the only senior manager I can discuss things openly with," and "When Winston was here, we were at least miserable together. Now we are miserable by ourselves." ("Miserable" here means working long hours.)

Lenovo's CEO, Young Yuanqing, received $3+ million more as a bonus in 2012 than in 2011. Instead of pocketing the extra cash, he divided it among the 10,000 lowest paid employees in the company. This surely indicates his *caring* for his staff. Young, a computer engineer by training, was promoted from an entry-level salesman to CEO in just 12 years, at the age of 37. He is no doubt a multi-talented person. But I have to assume that *his caring for his staff must have helped.*

Fast-Tracking Your Career: Soft Skills for Engineering and IT Professionals, First Edition. Wushow "Bill" Chou.
© 2013 The Institute of Electrical and Electronics Engineers, Inc. Published 2013 by John Wiley & Sons, Inc.

When Robert Rubin was the chief economic advisor for President Clinton, he kept $1 as his symbolic salary and divided his real salary among his staff. When he was the US Treasury Secretary, he told the staff to call him simply as "Bob," not "Mr. Secretary." He was amiable and unpretentious to the staff, and had a good rapport with his subordinates. He is a highly successful person and must be multitalented, but again, I have to assume that *his caring for his staff must have helped.* (In addition to the above, he has also held positions as chairman of the board at Goldman Sachs and at Citigroup.)

Practicing the following rules, defined later in the chapter, can help us achieve deftness in Motivating Smart:

- Winning loyalty by being loyal
- Getting credit by not taking credit
- Motivating by complimenting
- Never polishing a sneaker (Refer to Chapter 6, Career Smart, for a general discussion of this rule).

In addition, the success of this soft skill can be enhanced by the following soft skills:

- Communications Smart (Chapter 1)
- People Smart (Chapter 2).

RULE 1:   Winning loyalty by being loyal

Win genuine loyalty from our staff by demonstrating our loyalty to them. Loyalty from our staff must be earned; it cannot be obtained by demanding it.

Donald Regan, Treasury Secretary under President Ronald Reagan, once said, "You've got to give loyalty down if you want loyalty up." We can show our loyalty for our staff by standing up for them when they encounter undue difficulties, rewarding them adequately, and mentoring worthy staff. We want to show them that if they perform well, they can count on us to take good care of them. By doing so, we can expect to earn loyalty from our staff and expect them to be inspired, dedicated, creative, and highly productive.

EXAMPLE: Protecting staff against layoffs

A CFO/controller wanted to cut several employees to boost the company's net profit. He asked every VP to cut the same percentage of staff in each of their groups. All the VPs complied, except for one, Wilson, who fought tooth and nail against the decree. His argument was that employees should be laid off based on their performance and value to the company, not by an arbitrary percentage cut. This made sense because it was an accepted fact throughout the company that this VP's staff were all good performers and that they brought in the most revenue for the company. The CFO/controller was not deterred and responded by saying that there had to be some ranking within the group, and Wilson could still identify those who were not as productive as the others. Wilson retorted by saying, "Okay, I'll submit one name— mine—for not being able to protect good performers from being laid off arbitrarily and for not following your decree." In the end, Wilson won, and nobody in his group was laid off. Naturally, this act alone earned Wilson tremendous loyalty from his staff.

EXAMPLE: Protecting staff against false accusation

The government's working environment is more complicated and politically charged than others, so the example of a boss being loyal to his subordinates is much more noticeable there. An Assistant Secretary of Management (ASM), George, in a federal agency made a big impression by demonstrating loyalty to his staff.

Due to conflict of interests, someone tried to inflict harm on one of George's direct reports by falsely accusing him of doing something improper. The staff reported the false accusation to George immediately. Without pausing, George said immediately, "I trust you. I will back you up. Don't worry about it. Things like false accusations happen all the time." This was an act of staff loyalty. Of course, the staff was very much moved. This also had a positive morale boost for other direct reports as well.

EXAMPLE: Obtaining special awards for deserving staff

When one of his direct reports had to leave for another job, George (mentioned in the previous paragraph) successfully sought a meritorious service award for him. Although it indicated George's appreciation for a job well done to one person, it also showed other senior staff members that he knew how to show appreciation for jobs well done.

EXAMPLE: Sharing credit

Barry was a low-level programmer at a consulting firm. His main responsibility was to run a software system with different input/output data points and organize them into curves. Normally, people with positions like Barry's wouldn't be listed as coauthors on a report, but his manager, Andy, was different. Andy listed everybody who participated in the effort as a coauthor, including Barry. Barry was delighted and appreciative. (This was the first time Barry saw his name listed as an author.) Later, as Andy was putting together a paper for publication based on that initial report, he realized he needed more rigorous data to support the paper's conclusion. Barry willingly worked over the weekend on his own time to generate more data points, all because his name was listed as a coauthor. In this example, Andy sacrificed nothing by listing Barry as a coauthor, yet he got a dedicated worker who was willing to give up his weekend.

📋 **EXAMPLE: Disney-Pixar listing every technical staff in a movie's credit section**

If you've ever seen a Disney-Pixar-produced animation movie, you might have noticed the long list of credits at the end. This list is so long because it includes the names of everyone on Pixar's technical staff, regardless of whether those people actually worked on that particular movie. This inclusion costs Disney-Pixar practically nothing, but it gives the staff pride and public recognition and enhances their loyalty to the company.

RULE 2:   Getting credit by not taking credit

Refrain from taking direct credit from the staff's achievement. We will get the credit anyway.

The best way for a manger to get full credit for a job well done by his staff is for him not to claim any direct credit for himself. The more he publicizes the staff's achievement, the more he himself gets the credit.

When a staff performs a project successfully, his manager often gets the credit whether he claims it or not. On the flip side, if the staff does a poor job on a project, the manager also gets the blame. Therefore, if we are the managers, we might as well be magnanimous and give our staff as much credit as possible when things are successful and share some blame if things don't go well. We have lost nothing and gain appreciation and loyalty from our staff. (Oddly enough, some managers don't seem to understand this very simple fact and try to claim as much credit for themselves as possible and push all the blame to their staff. It backfires more often than not.)

📋 **EXAMPLE: Getting full credit without claiming it**

In the mid-1990s, a hot topic in several government agencies was IT enterprise architecture, but nobody seemed to know exactly what it was, let alone provide any documentation. The US Department of Treasury's CIO, Willie, tasked a senior staff, Samuel, with developing a document on the department's enterprise architecture framework. It was a big hit, and the Executive Office of the President mandated every cabinet agency to develop a similar document modeled after Samuel's work.

Whenever he could, Willie gave full credit to Samuel and never claimed any for himself. He repeatedly mentioning Samuel's name to his direct boss, GM, telling him that Samuel was the one who did all the work. But somehow, Samuel's effort and name never registered in GM's mind—instead, he gave Willie full credit for having the document developed. This was evident by the following event: when GM left Treasury and became the head of a government-operated corporation. He asked Willie to join him. Willie, due to personal reasons, had to decline. Instead, he recommended Samuel. GM's immediate response: "Does Samuel know architecture?"

RULE 3:   Motivating by complimenting

Motivate staff through genuine compliments and encouragement.

With few exceptions, it is easy to find something to compliment or at least to encourage. However, the effect of compliments and encouragements would be even more profound if staff are working on or have been assigned with tasks that can match their skills, or at least are within their capability.

Any compliment given must be sincere. An insincere one can backfire.

EXAMPLE: From a so-so worker to a star performer

Paul was a new COO at a company. During his first few days, he had a private face-to-face meeting with his senior staff. Randy was considered a mediocre worker, but during their conversation, Paul discovered that Randy was actually quite intelligent. He was very much interested in Web design and had good ideas on how to improve the website. (Apparently, Randy's direct manager had not properly appreciated Randy's skill and interest.) Paul complimented Rand for his insight and skill. The next day, Randy presented even more solid ideas. (Apparently, he did this at home on his own time.) At the time, the company's website design and maintenance were outsourced to a contractor. Paul decided to put Randy in charge of the website design, with the contractor providing the support. Randy immediately changed from a mediocre worker to a highly motivated and dedicated performer. He even worked on his sickbed while recuperating in the hospital after an operation. The company's Web appearance and content made a quantum jump in just a couple of months. When the COO left the company for a CEO position at another company, he took Randy with him.

CHAPTER TEN

Delegating Smart

Principle: Assigning responsibility judiciously

Strategy: Match the right people with the right responsibilities at the right time

Delegating is not just about distributing responsibilities. It is about matching the right person with the right responsibilities at the right time.

President Ronald Reagan once said, "Surround yourself with the best people you can find, delegate authority, and don't interfere as long as the policy you've decided upon is being carried out." We do not have President Reagan's luxury and resources in choosing our staff. Applying to us, "Surround yourself with the best people you can find" is equivalent to identifying skills and personalities of each of our staff; "delegate authority, and don't interfere" is equivalent to delegating the right person with the right responsibility; and "the policy you've decided upon is being carried out" is equivalent to ensuring the person being delegated with the responsibility has the same view and philosophy as we do.

Practicing the following rules, defined later in the chapter, can help us achieve deftness in Delegating Smart:

■ Getting more done by doing less

■ Delegating successfully by matching tasks with staff

■ Making controversial decisions by not making them.

Fast-Tracking Your Career: Soft Skills for Engineering and IT Professionals, First Edition.
Wushow "Bill" Chou.
© 2013 The Institute of Electrical and Electronics Engineers, Inc. Published 2013 by John Wiley & Sons, Inc.

In addition, the success of this soft skill can be enhanced by the following soft skills:

- Communications Smart (Chapter 1)
- People Smart (Chapter 2).

RULE 1: Getting more done by doing less

Avoid "micromanagement" because it achieves less from doing more.

Most micromanaging managers are smart and capable. For first-level managers, micromanagement can be a virtue—it often leads to excellent performance and thus earns the manager a fast promotion. Unfortunately, such managers usually can't give up micromanagement when they move to a higher-level position. Micromanagement ultimately distracts these senior managers from spending adequate time and energy on more important issues, and it demoralizes the staff under their charge. This situation epitomizes the Peter Principle (every employee tends to rise to his or her level of incompetence) and leads to a triple loss: a loss for the organization, a loss for the senior manager, and a loss for the staff.

EXAMPLE: Micromanagement hinders career advancement

Donald is a detailed, organized, and competent technical professional. He was an excellent first-level technical manager who knew each of his staff members' strengths and weaknesses and how each was progressing on projects. He was soon promoted to second-level manager, and although he then oversaw a much larger group, he could still micromanage the staff.

However, when promoted to vice president, his micromanaging style finally came apart at the seams: staff morale was low because his micromanagement caused delays and additional work. For example, business trips now needed his approval, which made the travel-request process longer. Also, his staff used to just informally report on project progress to their direct manager. Now, they had to give a formal presentation to the VP. Because of his micromanagement style, Donald couldn't find enough time for long-term strategic planning—a critical component of his position. He was relieved of the VP position a few months later.

EXAMPLE: Micromanagement wastes energy on trivial issues

Al was a VP responsible for providing administrative and managerial services for a company of 1500 employees. His responsibilities included high-budget items—computer-related services and systems, and building renovation and maintenance. He rarely, if ever, reported to his boss areas in which he could save the company money, but he once proudly reported that he had figured out how to reduce toilet paper waste. With tens of millions of budget dollars under his care, this "proud" savings was far below the noise level.

He should have focused on more important areas that could have saved the company millions of dollars while providing enhanced services. This case symbolizes the biggest problem with micromanagement—focusing on numerous trivial issues instead of fewer but far more important tasks. He did not stay on the job long.

RULE 2: Delegating successfully by matching tasks with staff

Understand the strength and interests of each of our direct reports, and assign tasks and delegate responsibilities accordingly.

Delegating is easy, but successful delegation is an art. We need to know when, what, and to whom to delegate—the last of which is particularly difficult. The people to whom we delegate tasks must be in sync with our philosophy, have the right skills to excel, and be motivated to succeed. In matching tasks with our staff or other individuals, keep in mind that in any profession, at almost any office, only a small percentage of people can carry heavy water. We need to identify these people for critical tasks or important responsibilities.

But we still need the other people, the majority, to carry out other tasks that are comparatively less challenging yet still essential. We need to keep in mind that *almost everyone is willing to take on responsibilities, because people want to feel important and to savor the success of a job well done. Matching the right task to the right person will have this effect.*

EXAMPLE: Making nonproductive staff productive

Sammy and Lance were both conscientious workers. But in spite of their education in the right fields, they did not have what it takes to keep up with advancement in technology and did not perform well in their respective groups. Normally, people like this are managers' headaches. In their cases, however, their managers did find the right tasks for them.

Sammy's manager assigned him to interface with all outsourced vendors. His technical background was adequate in understanding the deliverables, and his conscientiousness led him to monitor vendors' progress and quality of work meticulously.

Lance's manager assigned him to be in charge of training interns. His background was more than adequate for this task. Being conscientious, he was able to work out excellent programs for training the interns. Having an avuncular manner, he was excellent in building good relationship with the interns.

In both cases, the managers found the right assignments for two individuals, making them very productive.

🖥 **EXAMPLE: Corporate CIO benefits from delegating to division CIOs**

When Victor took a new job as a corporate-level CIO, he noticed that, in spite of monthly division CIO meetings, there wasn't much cooperation among CIOs. He reasoned that this was because the meetings were owned and managed by the corporate CIO and his staff. The corporate CIO chaired the monthly meeting, and his staff set the agenda and managed all of the action items. The division CIOs didn't have much chance to interact with each other. He further reasoned that division CIOs were typically leaders and more involved if they had leadership positions; otherwise, they were more likely to be passive.

Victor decided to change this. He started calling the monthly meeting the CIO Council and delegated the council's management and ownership to the division CIOs. A division CIO would be elected by his peers to chair the council. The chair would set the agenda and manage the action items. Staff at the corporate CIO's office would facilitate the action items but not own or manage them. To make the division CIOs feel more special, he mandated that seats around the conference table were reserved for CIOs only.

The division CIOs were delighted with the change. The ownership of the meeting gave them a psychological boost. More pragmatically, going to the meetings used to be a chore dictated by the corporate CIO and his staff, but now it was a platform for collegial discussions about issues of importance to the division CIOs—such as leveraging each others' expertise and resources and identifying joint projects. In short, by delegating the responsibilities at CIO meetings to division CIOs, the meetings became much more effective, productive, and meaningful, with the added benefit of freeing up some of the corporate CIO office's time and resources. It was a win-win arrangement.

🖥 **EXAMPLE: Negative consequence from delegating responsibility to the wrong person**

Tom was a newly appointed division director. Tom brought with him a new IT manager, Garry. Garry was very bright and did a good job at the post, except for one problem. Although he was very talented in IT matters, Garry was also very talented at rubbing people the wrong way. His belligerent personality made him quite unpopular. The division staff's unhappiness with Garry propagated to Tom, to the point where almost anything Tom did was interpreted negatively, and Tom became equally unpopular. Several managers and senior technical staff members joined forces and requested that the VP who hired Tom remove Tom from his post. The VP relented. I'm certain that had Tom picked a more amiable person for the IT manager position, he would not have been removed from his post by the VP.

📖 EXAMPLE: Dividing up responsibilities judiciously

People in charge of professional magazines and conferences are typically volunteers who contribute their time and talent for professional visibility. This undertaking is, in general, viewed as laborious and time-consuming. But it does not have to be so.

A conference chair or a magazine editor in chief should not consume their time on running. Instead, they should spend their time on managing, which includes identifying other willing volunteers, defining tasks, and matching the tasks with the right volunteers.

I've run a couple of conferences and magazines quite successfully, if I may say so myself, without having to devote too much of my own limited time and without having to depend on my limited capability. How? I realized that many professionals, I myself included, enjoy professional visibility. If I can help talented professionals with such visibility, I should be able to recruit enough people more capable than myself to help me out.

For conferences, I divided the overall effort into separate tasks and create a committee for each task; I then recruited an appropriate person as the "committee chair." (Note that the committee chair title sounds more impressive than just a conference committee member. Being chairs, they also could recruit additional committee members to help them.) I had every task done by others, but these tasks and committees still had to be coordinated. For this purpose, I created two more positions, the vice general chair and the floor manager, to be responsible for all the coordination. I considered my main responsibility as strategic planning: defining the tasks and recruiting the committee chairs and vice general chair. Once these were done, I did not have to spend much of my own time and just watched the conferences run successfully. Understandably, the actual tasks and the strategy of committee chair appointment may vary with conferences. To attract attendees when I was the general chair for an annual IEEE North Carolina Symposium, I added a job fair to the program and invited a highly regarded HR manager of a local company to be in charge of it. To attract attendees when I was the general chair for an international symposium held in Mexico, I invited a professor from a Mexican university as the vice chair and a Mexican government minister as the honorary general chair for the conference.

For magazines, I used in essence the same strategy. I created more editorial positions and invited people more talented than I am to handle the actual editorial work. When I was editor in chief (EIC) of the *Journal of Telecommunications*, I made most issues feature special themes on hot topics. For each of these special issues, I recruited a well-qualified individual as the guest editor to be in charge of it. Each issue was practically run by this highly qualified individual on the special issue topic. I did not have to do much editorial work, if at all. When I was EIC of *IT Professional*, I divided my responsibilities into three areas and created three associate EICs (AEICs), with each responsible for one. With them in charge, I did not have to do much in running the magazine, except for occasional strategic planning. The three AEICs I recruited are more capable than I am and could do a better job than I could anyway.

I am a little bit embarrassed to get credit for success of these conferences and magazines. If I may borrow President Reagan's quote again: "Surround yourself with the best people you can find, delegate authority, and don't interfere."

RULE 3:   Making controversial decisions by not making them

Create a study committee to make recommendations on contentious issues. By controlling the committee's composition, we can have the committee's recommendations sync with ours.

A manager can delegate most tasks to others but not critical decision-making. However, he can always set up committees to study complicated issues and make recommendations. This is a particularly useful strategy when making a controversial decision or setting policy that will upset some people no matter which option the manager chooses.

The trick is to ensure that the committee's recommendations meet our objectives. It helps to organize the committee in such a way that most of its members share our viewpoints or are loyal to us. We may now appear to make our decision by accepting the recommendations given by the committee. This usually minimizes the level of dissatisfaction stemming from the decision.

EXAMPLE: Create a committee in an academic environment

Mike was a very skillful academic administrator. University faculty can be an unruly bunch—they don't always respect authority—so successfully holding an academic administrative position requires being popular among the faculty. Of course, making controversial decisions comes with the risk of lowering an administrator's popularity. Mike had a way of using committees to mitigate this risk.

When tackling a controversial issue, Mike would talk to people to get ideas. He would then make up his mind about what he intended to do. Instead of informing people of his decision, he would form a committee. Since he had already chatted with people about the issues, he could appoint a sufficient number of committee members who he knew shared his viewpoints. The committee recommendation would then more than likely match what he already had in mind. However, seemingly making his decision based on the committee recommendation lessened the negative response. He was very successful in using this strategy.

🖥 EXAMPLE: Create a committee in a government environment

Typically, the Assistant Secretary for Management (ASM) at a federal cabinet agency is responsible for agency-wide administrative services and policies. One such function is resource allocation, which is always controversial. One ASM tried to minimize the controversy by forming a committee to make such decisions. He had four people from his office populate it, along with four representatives from other offices in the agency. By controlling the agenda, his staff had the advantage of influencing the other members, so the committee usually made recommendations that matched the ASM's interests.

However, politics being politics, sometimes the ASM's staff manipulated the agenda to generate their own preferred results and not necessarily those of the ASM. They could easily blame the other committee members for the outcome.

PART SIX

Being Visionary
LEADING TO THE C-SUITE

CHAPTER ELEVEN

Beyond the Box

Principle: Thinking like an entrepreneur and developing forward-looking visions

Strategy: Look for potential opportunities beyond the confines of our working environment

The common denominator of all extraordinarily successful people is a disposition toward thinking beyond the box and a will to pursue their vision. It is a three-step process: examine the bigger picture, form a visionary plan, and then execute or market the vision.

Conventional wisdom leads us to think that we're doing a great job if we've successfully performed our prescribed tasks. Indeed, we have. But *to fast track our career, we need to think and act like an entrepreneur.* We should not confine ourselves to the box as tasked. Instead, we need to think beyond the box, aspiring to enhance our current and future career.

Ever since the inception of the Internet, we have witnessed many, many successful entrepreneurs. They are not just the computer gurus: they include housewives and students, among others. *They all think like entrepreneurs and look for potential opportunities beyond the box, then pursue and market their vision with perseverance.*

In the "Boss Smart" section, I mentioned "farsighted" as a trait that any boss appreciates. I feel that this is an important topic and deserves more discussion because it isn't limited only to dealing with a boss. It could have a big impact on our whole career. *"Think beyond the Box" may be considered as being farsighted on steroids.*

Fast-Tracking Your Career: Soft Skills for Engineering and IT Professionals, First Edition.
Wushow "Bill" Chou.
© 2013 The Institute of Electrical and Electronics Engineers, Inc. Published 2013 by John Wiley & Sons, Inc.

Practicing the following rules, defined later in this chapter, can help us achieve deftness in thinking Beyond the Box:

- Examining the big picture to identify opportunities
- Forming a visionary plan
- Marketing the vision
- Being proactive and farsighted (Refer to Chapter 8, Boss Smart, for a general discussion of this rule.)
- Never polishing a tennis shoe (Refer to Chapter 6, Career Smart, for a general discussion of this rule).

In addition, success of thinking Beyond the Box can be enhanced by the following soft skills:

- Communications Smart (Chapter 1)
- People Smart (Chapter 2)
- Marketing Smart (Chapter 3)
- Work Smart (Chapter 4)
- Time Smart (Chapter 5)
- Boss Smart (Chapter 8)
- Motivating Smart (Chapter 9)
- Delegating Smart (Chapter 10).

RULE 1:   Examining the big picture to identify opportunities

Examine how changes can be made to enhance our or our group's performance, examine how we can affect the operations of others, and examine how others can impact us.

With respect to the overall organization, what we do is akin to being a branch on a tree in a forest. What happens to the branch can impact the tree, what happens to the tree can impact the forest, and what happens to the forest can impact the whole environment. And vice versa.

To ensure that what we do achieves its maximum impact, we need to examine and reflect upon the following.

Our own skill set

In today's environment, succeeding as a senior manager requires having a multitude of talents. Is our current skill set adequate? Do we need to enhance it? Have we kept up with relevant technology advancement and business changes?

For example, if we're a technical manager with a strong technical but weak managerial background, we might want to consider acquiring business skills to complement our technical expertise. Or, if we're in charge of operations, we might want to develop a working knowledge of technical design and product marketing.

Our staff's skill set

Nowadays, the skill sets we need from our staff are quite fluid. This is particularly so for technical staff. Does the staff have an adequate, up-to-date skill set?

For example, many IT shops have been comfortable outsourcing IT expertise. However, as needs change, we might need to consider acquiring or developing mission-critical expertise in-house. At a minimum, our staff should be fully capable of getting the best out of contractors. This could be as simple as obtaining a multimedia presentation software package or as complicated as developing data mining and security capabilities.

Operations

Can fine-tuning the functionality and operations under our control result in improving the effectiveness of our group? Can it help improve the operations of other groups in the organization?

For example, if we're running an IT shop that provides IT services, we might need to constantly focus on improving user friendliness and broadening application services.

Collaboration

Can our group and others form a symbiotic relationship? Groups working on similar types of projects and problems can benefit greatly from working together, saving money and improving performance. Furthermore, not working together can have costly consequences.

A notable example is the 2009 Christmas day shoe-bomber fiasco. The would-be bomber was on one intelligent agency's terrorist watch list but not on another agency's no-fly list. Consequently, he was allowed to board a plane flying from Amsterdam to Detroit.

These intelligence agencies are supposed to share information. Of course, had this really happened, there would have been no technical reason for not having prevented the would-be bomber getting on board. Luckily, the would-be bomber was incompetent and his fellow passengers vigilant; otherwise, the lack of an authentic symbiotic relationship between these agencies' IT groups could have resulted in a deadly disaster.

Hopefully, the involved agencies will now think beyond the box of their own turfs and form a much closer, more symbiotic relationship—particularly in their IT departments, where the payoff can be great and the impact to our national security significant.

Speaking of closer and symbiotic relationship in IT departments, this should not be limited only among intelligent agencies. Billions can be saved in federal government's annual IT budget and performance much improved if all Federal IT departments can leverage each other. It is too bad that nobody in the position to make it happen has had the will and the farsightedness to do so.

Leveraging technology

It's always good to ask if our group is taking full advantage of available technology to benefit the organization. Have we leveraged Web, software, and other technologies to enhance our mission?

Consider Web implementation. A thoroughly planned and properly implemented website can significantly enhance an organization's mission, increasing its sales and burnishing its image. Yet after over 20 years' existence, Web technology has yet to be adequately deployed by most organizations to make full use of its potential.

A good website should intuitively and promptly provide any information that the user seeks or that the organization wants to provide. A good website should provide tools to enhance performance of the organizations' staff. A good website should provide data that can enhance the organization's mission. If an organization's website doesn't live up to these objectives, its CIO is guilty of dereliction of duty.

Consider smart software: a thoroughly planned and implemented software system can significantly enhance an organization's mission. If we are willing to look outside the comfort zone, we can define smart software systems that could potentially do wonders! For example, a smart software system could link independent databases in real time to mine or deduce information that would otherwise be difficult to achieve. A smart software system could potentially coordinate Internet switches to block or reduce the damage of denial-of-service attacks. Unfortunately, this is one area whose full potential is not readily appreciated.

Changes in markets and technology

Is our group or are other groups in the organization ready to meet the challenge and opportunities available from technological advancement and market changes? This task is particularly relevant for CIOs. In most organizations, the CIO is, or should be, the senior manager most knowledgeable about technological advancement and its impact and should be the one to keep the CEO abreast of such issues.

It is well known that Wal-Mart's success is due in part to its deft application of state-of-the-art IT technology, infrastructure, logistic, distribution, and operations (http://www.cio.com/article/147005/45_Years_of_Wal_Mart_History_A_Technology_Time_Line).

While Wal-Mart is the poster boy for success in utilizing IT, the US Postal Service (USPS) exemplifies an organization that appears to have failed in making good use of IT. (The following scenario is my personal take and opinion. Others may partially or totally disagree.)

The USPS has been, for more than 40 years, unable to think beyond the box it has been in for more than 200 years. In spite of its superbly advantaged infrastructure, it has lost business to FedEx, UPS, and the Internet. It has lost money, big time. Cutting delivery on Saturday and raising postage rates won't solve its problems in the long run and might actually backfire, hurting its business.

In recent years, the USPS has shown signs of thinking beyond the box. If the USPS could think more like entrepreneurs, this service could conceivably exploit its infrastructure superiority and regain some, if not most, of the business it has lost to FedEx and UPS. It could combine its infrastructure strength with Internet technology to generate value-added new revenue.

RULE 2:   Forming a visionary plan

Form a vision and map out an action plan for realizing it.

Once we've identified actions to perform, we need to identify visionary objectives that are worth pursuing.

For example, consider the electronic medical record (EMR) system. A typical EMR system designed for physicians may provide the following objectives: clinical notes and patient health and treatment data, prescription management, and access to a drug and lab order database. But in continuing to think beyond the box, a more visionary EMR system could be expanded to contain the following additional functions: the ability to provide an analysis, suggest actions, or assist physicians with diagnosis using an expert system. It should have a function that allows doctors located at different sites who have to treat/diagnose the same patient to have the same access to that patient's database. I am amazed why so few people have the vision to build such a system. It could substantially improve patient care, reduce medical costs, and even save lives.

An EMR system designed for patients could also give them access to their Personal Health Records (PHR). A typical PHR would offer access to test results and pending appointments. However, a more visionary PHR could add access to clinical notes, detailed tutorial information related to the patient's illness, and relevant information on prescribed medications.

Additional examples appear later in the chapter.

RULE 3:   Marketing the vision

Articulate the value of our vision to stakeholders if the action plan for realizing the vision is outside our control and authority.

Some visions can be implemented or executed under our own volition or authority. For example, under Rule 2, Chapter 10, I mentioned a situation in which the corporate CIO wanted to establish a CIO Council, consisted of division CIOs. This was a change in *operation* that the corporate CIO could easily implement *on his own authority*. Another example to be given in the next section is about an SVP who realized the need to broaden his *skill set*. This of course was achieved under *his own volition*.

But some visions have to be supported by stakeholders. For example, under Rule 4, Chapter 8, I mentioned an individual named Henry who had a vision of adding high-speed products to his company's offerings. This would have been far beyond his authority, so he had to market the idea by making multiple presentations to his senior management before he could successfully *market his vision*. Another example in the next section is about a chair of the electrical engineering department at a university. He had the vision that computer engineering faculty in his department should work closely with computer science department faculty in developing graduate programs and in pursuing joint research projects. This vision would have been far beyond his authority. It took more than 1 year before he was able to convince the university administration as well as other stakeholders. But at the end, his vision was successfully marketed and realized.

I first addressed this topic in the "Marketing Smart" section, where I discussed how to market an existing or soon-to-be-completed product; here, I'm talking about marketing ideas for potential benefits in the future. This is akin to asking a venture capitalist to make an investment. A venture capitalist once told me "venture capitalists only invest in dreams." So, when marketing a vision, we need to paint as rosy a picture as we reasonably can.

SUCCESSFUL FAST-TRACKING STORIES

While highlighting the impact vision has had on the following individuals' careers, it goes without saying that they are talented on many fronts. Being visionary is just one of them.

🖳 EXAMPLE: From manager to director to vice president to senior vice president in 8 years

When Ron first received his PhD from a top research university, he did what a typical PhD graduate from that school would do—join an R&D group in a well-known organization. While he did well and learned a lot, he realized that to stand out among a group of top-notch researchers would be a high order. After having worked on that job for 2 years, he decided he had to "*hop to a pond*" where he could be a big fish.

He joined a production group at a large fabless electronics company as a first-level manager. (A fabless company designs special-purpose chips but doesn't manufacture them. Instead, it sends the design specs to a chip production manufacturer. In a fabless company, the production group is responsible for finding manufacturers that can cost-effectively produce the chips.)

Ron could have stayed "within the box" by simply finding reliable and reasonably priced manufacturers. However, he had a vision. He felt that to do an outstanding job as a product manager, he needed a skill set beyond that of a typical director of production. He believed that he needed a skill set of marketing, design, and manufacturing. (That was his *vision*.) He wanted to (1) develop marketing skills so that he could understand what features current and potential clients would need and desire; (2) have a good grasp of the design process so that he could provide the design engineers with pragmatic suggestions that the clients would desire; and (3) have a good understanding of the manufacturing process to put him in a better position to negotiate prices with manufacturers and to ensure good quality control. (That was his *visionary plan*.) The skill set he acquired gave enough know-how for him to do an outstanding job. His "beyond the box" vision was achieved under *his own volition*. But his resulting outstanding performance was recognized by the executive vice president. He was promoted to the position of director and then vice president of production in less than 4 years.

About a year after he became VP, we had dinner together, during which he had to respond to a Blackberry email message. He apologized and boastfully complained that, in addition to the VP position, he was also the general manager of a product line. When I saw him about 6 months later, he walked in while talking on his iPhone. He apologized again, this time boastfully complaining that in addition to the VP position, he was now the general manager of three product lines.

At that stage, he became too big a fish for the pond. The organization did not have a senior vice president (SVP) position, and the president and executive vice president were cofounders of the company. In other words, he didn't have room to move up, so he decided it was time for him to "*pond hop*" again.

He is now an SVP of a company that has a $16B annual revenue and is about four to five times bigger than the previous company he worked for. This company has more room for him to grow. As of this writing, he just began his new job, but I am certain that, with his visionary talent and skill, he will soon be promoted to even more senior positions.

Apparently, "looking beyond the box" paid off handsomely for him. Had he not done so, he probably wouldn't be living in his multimillion-dollar mansion today.

EXAMPLE: From senior technical staff to director to vice president in 4 years; to full professor in 3 more years

A professor from a leading research university decided to start a think-tank consulting and software development company. He started the company with two colleagues, one previous student and one recently graduated student, William. The faculty and students at that university have a tendency to be snobby at nitty-gritty real-world problems and a tendency to solve problems with mathematical rigorousness. While all five of them realized that they were not dealing with theoretic problems anymore and needed to readjust their attitude and solve real-world problems with different approaches, William took the most aggressive steps in this new environment. Here was his *visionary plan*: (1) learn details about nitty-gritty real-world problems, which would not be necessary when dealing theoretic problems; (*Pond happing to a more promising project; Showing enthusiasm for challenging assignments*) (2) be willing to make approximations and compromises in solving problems, which are different from the rigorousness normally required by theoretic problems; (*Achieving outstanding results by not seeking perfection*) (3) give speeches at conferences as a subtle way to market his and the company's expertise; (*Killing two birds with one stone*) and (4) publish papers to establish himself and the company as the authority on these types of problems (*Killing two birds with one stone*). Under this strategy, he performed

outstandingly well on whatever projects he was in charge of, and his importance to the company continued to grow.

In addition to William's performance, the company's president was
particularly impressed with William's loyalty. The president noticed that
whatever action William took, he always insisted on what was best for the
company. If they had any disagreement, William always tried his best to carry
out the president's decision in spite of his disagreement.

With the combination of his *visionary strategy* and his recognized *loyalty*
to the company and to the company president, William was promoted from
senior technical staff to director and finally to vice president in a time span
of 4 years; established a group that as a whole is the most capable and profit-
able in the company; and established an international reputation in his field
of specialty.

After he had worked at the company for 7 years, there was apparently
no room for him to grow. He decided to "*pond hop*" to a university. Normally,
it takes 10 or more years with productive research and teaching to get promoted to a full professorship. But with William's strategy of speaking and
publishing, he had built a good record of publications. He got an appointment as a full professor.

EXAMPLE: From department chair to dean to university chancellor

Lloyd was the department chair of electrical engineering at a state university
in the mid-1970s. In the computer field, he noticed the following: (1) an
overlap between courses taught by the faculty in the computer science department and those taught by his own department's computer engineering
faculty; and (2) years ago, the state-wide university system had decided to
implement only an undergraduate computer science/engineering program at
Lloyd's campus and only a graduate computer science/engineering program
at a sister campus.

Lloyd examined the big picture. He knew that it was important for the
university's future growth to have a strong computer science and engineering
PhD program. He formed a vision: (1) a strong computer science/engineer-
ing graduate program has to be established on his campus; and (2) the
computer science and computer engineering departments should leverage
each other's resources.

He faced two challenges: (1) how to prevent the objection from the sister
campus that was designated with the computer science and engineering
graduate program; and (2) how to get faculty of the two departments to work
together.

His *visionary plan* was to establish an ostensibly interdisciplinary
computer-based graduate program that he called the "Computer Studies
Program." He enlisted faculty from other departments on campus to participate: officially, it wasn't a graduate program for computer science or

computer engineering, but an interdisciplinary graduate program that contained computer science or computer engineering as part of the curriculum. With this argument, he marketed his visionary plan by fending off objections from the sister campus and convincing the university administration to approve and support the program. In reality, it was a graduate program with a joint effort from faculty in both computer science and computer engineering, and he convinced the two faculty that it was in their best interests to work together. To avoid potential leadership conflict, he found a program director from off the campus.

His effort paid off, and the Computer Studies Program was a success. Shortly afterward, he was made Dean of Engineering. (He moved from being a department chair to a dean in 4 years.) Of course, his ascension to dean depended on many of his talents, but being visionary on this and other projects obviously played a role in his appointment.

As Dean of Engineering, he continued to think beyond the box, including integrating academic programs and establishing new research centers. Under his stewardship, national ranking of the engineering programs at the university rose rapidly, putting it in the top 20. He was eventually made university chancellor.

Shortly after he was appointed as chancellor, the university's athletic department was in turmoil, and the athletic director resigned. Lloyd knew he could not find a top-notch athletic director who would be willing to take the position under the mess, so he did something *outside of the box*. He appointed an academic administrator to be the acting athletic director to straighten out the mess first. His strategy worked. With that done, Lloyd was able to hire an able athletic administrator as the permanent athletic director.

As chancellor, he envisioned a second campus where researchers from industry and the university could cross-fertilize. *He successfully sold this vision to the then governor*, who allocated a big piece of choice land to the university as its research campus. *Marketing the vision successfully to the endowment foundation*, he was able to raise substantial donations needed to build facilities on the new campus. This is by far the most impressive legend any university chancellor has ever achieved at this campus.

Lloyd is now retired. No one has been able to hold the chancellor position as long as he did. To be sure, his quick move up and his success on the career ladder were not just based on his ability to look beyond the box—he had many other managerial talents. Nonetheless, his visionary talent helped him greatly.

Final Thoughts

THE BOOK'S OBJECTIVE

Many of us have heard variants of the following motto, "if we are competent, work hard and complete our assignments, we will be properly rewarded." This is true, but only partially. It is a misguided myth. In reality, in order to be fully rewarded, our competency, our hard work, and/or completing our assignment must be appropriately complemented with "soft skills."

I have observed that, partly due to this misguided motto, many technical professionals are not cognizant of the importance of, not deft with, and/or not willing to learn, soft skills. I have also observed that technical professionals, if desired, can easily learn to be deft with soft skills. These observations motivated me to write this book. *I want to show engineering and IT professionals what the smart soft skills are and how these skills can help them fast-track their careers and assist them in reaching their career goals.*

"SOFT SKILLS" AND "RULES" OUTSIDE THE SCOPE OF THIS BOOK

If we are to ask different persons about what are the soft skills needed for fast-tracking to a successful career, we are bound to get different sets of answers. I wanted to be sure that the scope of this book cover all significant soft skill rules that are essential to fast-tracking our career. I sent the book's original draft copy to many qualified individuals seeking their advice on possible missing topics. I am very grateful that they suggested several rules for considerations. It turned out that with exception of one rule, all other suggested rules are either already covered in the book under different titles or are not relevant. (The exception: *being heard by listening*, suggested to me by Frank Ferrante.) These rules can be classified into five categories. Readers who are exposed with soft skill rules elsewhere will likely find they would fit into one of the following.

Fast-Tracking Your Career: Soft Skills for Engineering and IT Professionals, First Edition. Wushow "Bill" Chou.
© 2013 The Institute of Electrical and Electronics Engineers, Inc. Published 2013 by John Wiley & Sons, Inc.

Born with personality

The most notable beneficial personality that many may be born with is generally known as having "charisma," or being naturally "amiable" or "likable." I consider these traits as being something we are born with; they cannot be learned as soft skill rules. However, if anyone can master most of the rules as defined in this book for "People Smart," "Boss Smart," and "Staff Smart," she or he will effectively have the same benefits as being born with charisma, and then some. (Later in the section, I shall lump the personality and soft skills together and define the combination as "amiable persona or equivalent.")

Rules that have already been covered in the book more effectively

Some rules may have a slight variation in their titles but they have the same objective as rules that have been covered in the book. One such rule that has been suggested for considerations is "joining trade groups." This suggestion relates to networking. But the book already has a section about the best ways to successfully establish networking, including a recommendation to attend trade conferences and join professional groups.

Rules that are the effects of soft skills but are not soft skills themselves

Some rules may have been considered as soft skills, but I feel they are actually not. One example in this category is "finding and utilizing a mentor." If we can demonstrate our own professional and personal skills, we do not have to be concerned about how to utilize a mentor. Mentors will find us.

Rules that are mottos but not themselves soft skills

The most noted mottos are "professionalism" and "teamwork." They should be expected of all office workers. They are not specifically aimed at career fast-tracking and are not specifically soft skills.

Rules that are not essential to career fast-tracking

Some soft skill rules are very important for various purposes but are not essential to career fast-tracking. For example, "how to fire staff skillfully" is an important soft skill for a manager. But I do not think it is essential to career fast-tracking. Another example is hiring. I actually wrote an article on smart hiring. I have not included that article's material in this book, because while hiring is a very important managerial skill, I do not consider it essential to career fast-tracking.

Rules that are expected of any competent professional

There are certain rules that are expected to be the norm for every professional. These rules should be followed regardless of whether one is interested in fast-tracking one's career. One example is replying to emails within a reasonable timeframe. (There is no specific guideline. But I usually respond within 24 hours of its receipt.) Another example is an interesting story told to me by a US expat working for Samsung in Korea. Over there, when a person with a more senior rank arrives at the office in the morning or leaves the office at night, everybody with a lower rank stands up.

HIGH ACHIEVERS' SOFT SKILLS

I can quite confidently state that almost all career high achievers are good with almost all the soft skills as I presented in the book. However, it does not mean they excel in all of them. They generally excel at just a few. My observation of 12 high achievers (CEOs, senior executives, university presidents, and members of National Academy of Engineering) has led to my identifying the soft skills they specifically excel at. The result is listed in the following table.

| | Good in Most Soft Skills in Book | Excel in Amiable Persona or Equiv. | Excel in Thinking Beyond the Box | Excel in Comm. | Excel in Pond Hopping | Professional Skills | |
						Very Good	Excel
CEO 1	x	x	x	x		x	
CEO 2	x		x	x			x
CEO 3	x	x	x	x		x	
SVP	x	x	x	x		x	
Sen. Ex 1	x	x	x	x	x		
Sen. Ex 2	x	x	x	x		x	
Sen. Ex. 3	x	x	x	x		x	
U. Pres. 1	x	x	x	x		x	
U. Pres. 2	x	x	x	x		x	
U. Pres. 3	x	x	x	x		x	
Acad. Eng. 1	x	x	x	x			x
Acad. Eng. 2	x	x	x	x			x

Sen. Ex., senior executive; U. Pres., university president; Acad. Eng., member of National Academy of Engineering.

My observations, as shown in the table, tell me that with the exception of CEO 2, every one of the high achievers excels in personal charisma or equivalent. But the CEO 2 compensates his lack of personal charm with his excellent professional expertise. Another exception is Sen. Ex. 1. Normally, it helps for an executive to be very good at his professional field, and Sen. Ex. 1 is not (though he is still quite good). He compensates this with his excellent skills in pond hopping.

My observation leads me to make the following conclusion. If we aspire to be a high achiever in our career, we need to be good in most, if not all, of the soft skill rules discussed in this book. We need to excel in soft skill rules associated with amiable persona, communications, and thinking beyond the box, and to be very good at the hard (professional) skill of our field of specialty. If we are a little weak in the soft skill area, we must really excel in the hard (professional) skill. If we are good, but not very good in our professional skills, we must be stronger in soft skill area than others.

PERSONAL CAREER GOALS

For personal reasons, different persons have different career goals. Some aspire for the C-suite, some not. Some strive for the managerial ladder, some are happy staying as a technical staff.

Andy Grove resigned as CEO from Intel in 1998, a few years after having been diagnosed with, and successfully treated for, prostate cancer, while Steve Jobs continued as Apple's CEO until his death even though he had been diagnosed with a much serious cancer and been treated unsuccessfully. Warren Buffet, 82, as of this writing, is still active as CEO of Berkshire Hathaway, while Bill Gates resigned as CEO from Microsoft in 2000 at the age of 45.

Ken, a senior director as his company, has a fairly good chance to move up to a VP position if he stops his weekly commuting between where he works in Washington and where his family is in California, or if he is willing to move his family from California to Washington. On the other hand, Jim, the chief scientist of a large R&D organization, moved to the present position in part because he was willing to move his family with him three times during the last 10 years.

Frank, previously a senior manager, was not interested in moving up the corporate ladder, because he was concerned that he would not have enough time to spend with his children and grandchildren. On the other hand, Ron, an SVP, has to travel frequently away from his children in moving up to his current position.

As I was leaving the US Treasury, I was offered the opportunities to advance my career then and in the future. Due to health reasons, I declined these opportunities. On the other hand, a colleague of mine at the time, actively pursuing advancement in his career, is now a US cabinet deputy secretary.

Our personal goals may be different. However, soft skills as espoused in this book are equally important. Even though the conclusion I outlined in the last section was drawn from my observation of high achievers, it is equally applicable to all of us, regardless of our career goals.

Appendix Tables for Principles, Strategies, and Rules

TABLE A.1 Principles and Strategies

	Basic Principle	**Basic Strategy**
Communications smart	Staying succinct and focused	Get key points across within the audience's limited attention span
People smart	Making people feel good	Put ourselves in other people's shoes
Marketing smart	Striking a chord with "customers"	Promote our "product" (ourselves, our work, or our charge) in its best light
Work smart	Focusing on best return with reasonable effort	Seek out methodologies that can bring best returns from our efforts; reject, give up, or adjust tasks that do not otherwise get good returns
Time smart	Recognizing time as more valuable than money	Aim for good ROI; turn spare time into opportunities; minimize nonproductive time
Career smart	"Pond hopping" strategically	Find organizations or positions in which our strength gives us the advantage and opportunity
Job-interview smart	Striking a chord with interviewers	Prepare well and conscientiously; stay away from any attitude that could be construed as cavalier or arrogant
Boss smart	Making the boss look good	Allow our boss to claim credit and benefit for what we have done, promote his/her policy, and shield his/her weaknesses
Motivating smart	Caring for staff	Keep in mind the best interest of staff's career and act accordingly
Delegating smart	Assigning responsibility judiciously	Match the right people with the right responsibilities at the right time
Beyond the box	Thinking like an entrepreneur and developing forward-looking visions	Look for potential opportunities beyond the confines of our working environment

Fast-Tracking Your Career: Soft Skills for Engineering and IT Professionals, First Edition.
Wushow "Bill" Chou.
© 2013 The Institute of Electrical and Electronics Engineers, Inc. Published 2013 by John Wiley & Sons, Inc.

TABLE A.2 Communications Smart

Principle: Staying succinct and focused
Strategy: Get key points "across" within the audience's short attention span

Rules	Explanation
Being always ready for elevator pitches/speeches	Be prepared to articulate short pitches at any brief opportune encounter in order to make the right impressions with the right people.
Mastering a presentation by mastering the onset	Summarize key points at the onset of a presentation or a written document
Using three diagrams to simplify complexity	Illustrations help explain complicated problems, but try to limit them to no more than three
Sizing up and resonating with the audience	Assess the audience's interests and background, then fine-tune a presentation to resonate with the audience (keep in mind that different people may resonate differently)
Being careful of careless comments	Be mindful of what we say. Carelessly worded comments can lead to negative reactions
Using plain language	Use words and expressions that our intended audience can readily understand
Using jokes and self-deprecating humor	Have a sense of humor and humility by poking fun at ourselves and mixing jokes with our presentations (It helps to draw audience attention to our presentations)
Being heard by listening (Refer to Chapter 2, People Smart, for a general discussion of this rule.)	Be attentive to what others have to say: by doing so, we learn the speaker's viewpoint, and are better prepared to communicate our viewpoint
Making a convincing presentation by making a well-crafted presentation (Refer to Chapter 3, Marketing Smart, for a general discussion of this rule.)	Craft presentations to convey our positive spin and deliver information that can resonate with our audience

Please refer to Chapter 1 for more details.

TABLE A.3   People Smart

Principle: Making people feel good
Strategy: Put ourselves in other people's shoes

Rules	Explanation
Getting accepted by accepting others first	Accept people around us by assimilating into their culture. This is the easiest way to build rapport with people around us and to be accepted
Winning by understanding both ourselves and our counterparts	Knowing our own (and others') strengths, weaknesses, likes, and dislikes gives us the most advantageous position for mapping out winning strategies
Being aggressive by being nonaggressive	Maintain an outwardly nonaggressive attitude even if we mean to be aggressive in pursuing our career advancement (We can achieve more with less aggressive behavior)
Gaining by giving	Harvest professional assistance from our friends by doling out favor upon them first
Successful networking by networking less	Establish a successful network by building upon our visibility and credibility in our professional community (successful networking is measured by how we are known and by whom, not by how many people we know)
Being heard by listening	Be attentive to what others have to say: by doing so, we show respect to the speaker, learn the speaker's viewpoint, and are better prepared to respond
Getting liked by displaying self-deprecating humor in conversations (Refer to Chapter 1, Communications Smart, for a general discussion of "self-deprecating.")	Have a sense of humor and humility by poking fun at ourselves, which can make us more likable and project an image of being successful and confident
Avoiding gaffers by avoiding overconfidence (Refer to Chapter 4, Work Smart, and Chapter 7, Job-Interview Smart, for a general discussion of "overconfidence.")	Be vigilant not to step from confidence into overconfidence (which may unknowingly lead to arrogance, which may in turn result in negative reactions from people we encounter)

Please refer to Chapter 2 for more details.
Note: The success of this soft skill can be enhanced by the following soft skill: Communications Smart (Chapter 1).

TABLE A.4   Marketing Smart

Principle: Striking a chord with "customers"

Strategy: Promote our "product" (ourselves, our work, or our charge) in its best light

Rules	Explanation
Sizing up and resonating with our "customers"	Design and/or present our product's features in a way that resonates with stakeholders' needs, interests, and expectations
Putting a positive spin on our "product"	Put a positive spin on the strengths and weaknesses of the product to place it in its best light (this should be done without being deceptive or dishonest)
Making a convincing presentation with a well-crafted presentation	Craft presentations to convey our positive spin and deliver information that can resonate with our customers
Inciting enthusiasm with enthusiasm	Present our product with outward excitement, confidence, and/or enthusiasm (this can help gain customers' trust and spark their enthusiasm)
Being careful of careless comments (Refer to Chapter 1, Communications Smart, for a general discussion of this rule.)	Be vigilant not to make unguarded remarks (such comments could result in unexpected negative reactions from our intended customers)
Using plain language (Refer to Chapter 1, Communications Smart, for a general discussion of this rule.)	Use words and expressions that our intended customers/audience can readily understand

Please refer to Chapter 3 for more details.
Note: The success of this soft skill can be enhanced by the following soft skills: Communications Smart (Chapter 1) and People Smart (Chapter 2).

TABLE A.5   Work Smart

Principle: Focusing on the best return with reasonable effort

Strategy: Seek out methodologies that can bring optimal returns from our efforts; reject, give up, or adjust tasks that do not otherwise net good returns

Rules	Explanation
Achieving outstanding results by not seeking perfection	Refrain from being a perfectionist; instead, aim to achieve the best ROI (A zeal for perfection tends to draw more resources, but ends up with a lesser result)
Avoiding blunders of overconfidence	Be vigilant not to step from confidence into overconfidence (which tends to underestimate the amount of resources and efforts needed for success and ultimately ends in lesser results, if not total failure)
Focusing on self-examination, not on blaming others, when things gone awry	Focus on ourselves in mapping out a corrective path when things do not go our way (Pointing fingers at others does not in general help matters and can even be counterproductive)
Killing two birds with one stone (Refer to Chapter 5, Time Smart, for a general discussion of this rule.)	Plan to make one deliverable useful for multiple tasks (when feasible)
Never polishing a sneaker (Refer to Chapter 6, Career Smart, for a general discussion of this rule.)	Give up on projects/tasks that are unlikely to progress the way we had hoped them to

Please refer to Chapter 4 for more details.

TABLE A.6   Time Smart

Principle: Recognizing time as more valuable than money

Strategy: Aim for good ROI; turn spare time into opportunities; minimize nonproductive time

Rules	Explanation
Investing time with the same zeal as venture capitalists investing money	Invest intelligently and sufficiently at the onset of a project to enhance the chances of gaining a huge ROI at the end (this is typically venture capitalists' investment strategy)
Killing two birds with one stone	Plan to make one deliverable useful for multiple tasks (when feasible)
Minding ROI	Given a choice, opt for the task with better ROI; given an opportunity, enhance a task's ROI
Making nonproductive time productive	Do something productive when attending unavoidable, not fully productive activities (when feasible and appropriate)
Turning spare time into opportunities	Use spare time to engage in activities that could enhance our career
Keeping the mind sharp by taking catnaps	Take a short catnap at an appropriate time during the day, but only if feasible (a short nap can improve productivity)

Please refer to Chapter 5 for more details.

TABLE A.7   Career Smart

Principle: "Pond hopping" strategically

Strategy: Find organizations and positions in which our strength gives us the advantage and opportunity

Rules	Explanation
Opting to be a big fish in a small pond	Seek to work in an environment where we can stand out (It gives us a better chance to grow in our career.)
Hopping to a more opportune pond at opportune moments	Having achieved a certain objective at one organization, move to another one for a more promising career opportunity
Never polishing a sneaker	Move on to another organization/position, or give up in seeking career advancement, if we are unlikely to advance as we had hoped at our current position
Making a good lasting impression by making a good first one	Make a good first impression by performing extraordinarily well at the onset of a new position (the first impression is the lasting one)

Please refer to Chapter 6 for more details.

TABLE A.8 Job-Interview Smart

Principle: Striking a chord with interviewers

Strategy: Prepare well and conscientiously; stay away from any attitude that could be construed as cavalier or arrogant

Rules	Explanation
Being well prepared by collecting relevant information	Collect and study information relevant to the position, including details about the job, the organizational mission, and, if possible, the background of the interviewers
Putting a positive spin on our qualifications	Give ourselves a positive spin by playing up our strengths and playing down our weaknesses, especially with respect to the position we are applying
Preparing targeted elevator pitches/ speeches	Prepare a set of "elevator speeches" specific to anticipated questions that offer the opportunity to put a positive spin on our qualifications or to resonate with the interviewer
Sizing up and resonating with the interviewer	Tailor our statement to the interviewer's interest and expectation with the objective of inciting resonance with him/her (even more effective if the statement relates to our vision on how to carry out our responsibility)
Winning interviewers' confidence in us by exhibiting confidence	Present ourselves with confidence, and we will gain confidence from interviewers
Avoiding gaffes by avoiding overconfidence	Be vigilant not to step from confidence into overconfidence; it can lead to arrogance and gaffes, which may result in negative reactions
Being careful of careless comments (Refer to Chapter 1, Communications Smart, for a general discussion of this rule.)	Be vigilant not to make unguarded remarks (such comments could result in unexpected negative reactions from interviewers)
Using plain language (Refer to Chapter 1, Communications Smart, for a general discussion of this rule.)	Use words and expressions that the interviewer can readily understand
Inciting enthusiasm with enthusiasm (Refer to Chapter 3, Marketing Smart, for a general discussion of showing enthusiasm.)	Show enthusiasm with respect to the job we are applying for (this can help promote a positive impression with the interviewers)
Making a convincing presentation by making a well-crafted presentation (Refer to Chapter 3, Marketing Smart, for a general discussion of well-crafted presentation.)	If a formal presentation is required, create one that puts a positive spin on our qualifications and resonates with the interviewers

Please refer to Chapter 7 for more details.

Note: The success of this soft skill can be enhanced by the following soft skills: Communications Smart (Chapter 1), People Smart (Chapter 2), and Marketing Smart (Chapter 6).

TABLE A.9   Boss Smart

Principle: Making the boss look good

Strategy: Allow our boss to take credit and benefit from what we have done; promote his/her policy, and shield his/her weaknesses

Rules	Explanation
Winning trust by showing loyalty	Use actions and words that exhibit our loyalty to our bosses (but subtly and within reason)
Gaining gratitude by sharing credit and taking blame	Share credit with our boss for projects we have successfully completed and take the blame for what the boss has contributed that failed
Being astute by watching for nuances	Be astute in detecting nuances in our boss's expressions and "between the lines." From nuances, we can determine his real likes and dislikes and what he really wants done or not done.
Being proactive and farsighted	Actively look for ways to improve performance, expand services, and increase ROI
Showing enthusiasm for challenging assignments	Be willing to accept challenging assignments from our boss, as long as the required skills fall within the radius of our expertise (this can help promote a positive impression with our boss)
Being careful of careless comments (Refer to Chapter 1, Communications Smart, for a general discussion of this rule.)	Be vigilant not to make unguarded remarks (such comments could result in unexpected negative reactions)
Using plain language (Refer to Chapter 1, Communications Smart, for a general discussion of this rule.)	Use words and expressions that our boss can readily understand
Never polishing a sneaker (Refer to Chapter 6, Career Smart, for a general discussion of this rule.)	Move on to another organization or position, if our boss is inept or unreasonable

Please refer to Chapter 8 for more details.

Note: The success of this soft skill can be enhanced by the following soft skills: Communications Smart (Chapter 1), People Smart (Chapter 2), Motivating Smart (Chapter 9), and Delegating Smart (Chapter 10).

TABLE A.10   Motivating Smart

Principle: Caring for our staff
Strategy: Keep in mind the best interest of staff's career and act accordingly

Rules	Explanation
Winning loyalty by being loyal	Win genuine loyalty from our staff by demonstrating our loyalty to them in watching over and protecting their interests
Getting credit by not taking credit	Refrain from taking direct credit from our staff's achievement (by doing so, we may even get more credit)
Motivating by complimenting	Motivate staff through genuine compliments and encouragements
Never polishing a sneaker (Refer to Chapter 6, Career Smart, for a general discussion of this rule.)	Give up on those people who are unlikely to progress the way we had hoped them to

Please refer to Chapter 9 for more details.
Note: The success of this soft skill can be enhanced by the following soft skills: Communications Smart (Chapter 1) and People Smart (Chapter 2).

TABLE A.11   Delegating Smart

Principle: Assigning responsibility judiciously
Strategy: Match the right people with the right responsibilities at the right time

Rules	Explanation
Getting more done by doing less	Avoid micromanagement because it achieves less from doing more
Delegating successfully by matching tasks with staff	Understand each of our direct reports' strengths and interests, and assign tasks and delegate responsibilities accordingly (delegating is not simply redistributing responsibilities)
Making controversial decisions by not making them	Create a study committee to make recommendations on contentious issues. By controlling the committee's composition, we can have the committee's recommendations sync with ours.

Please refer to Chapter 10 for more details.
Note: The success of this soft skill can be enhanced by the following soft skills: Communications Smart (Chapter 1) and People Smart (Chapter 2).

TABLE A.12   Beyond the Box

Principle: Thinking like an entrepreneur and developing forward-looking visions

Strategy: Look for potential opportunities beyond the confines of our working environment

Rules	Explanation
Examining the big picture to identify opportunities	Examine how changes can be made to enhance our or our group's performance, examine how we can affect the operations of others, and examine how others can impact us.
Forming a visionary plan	Determine what we want to achieve and map out an action plan for realizing it
Marketing the vision	Articulate the value of our vision to stakeholders (not necessary if the action plan for realizing the vision is within our direct control)
Being proactive and farsighted (Refer to Chapter 8, Boss Smart, for a general discussion of this rule.)	Actively look for ways to improve performance, expand services, and increase ROI
Never polishing a sneaker (Refer to Chapter 6, Career Smart, for a general discussion of this rule.)	Give up on those visions that are unlikely to progress the way we had hoped them to

Please refer to Chapter 11 for more details.

Note: This soft skill can be further enhanced by the following soft skills: Communications Smart (Chapter 1), People Smart (Chapter 2), Marketing Smart (Chapter 3), Work Smart (Chapter 4), Time Smart (Chapter 5), Boss Smart (Chapter 8), and Delegating Smart (Chapter 10).

Abbreviations

CEO	Chief Executive Officer
CFO	Chief Financial Officer
CIO	Chief Information Officer. A large organization may have two or more layers of CIOs, with each administrative unit having its own CIO or IT manager. The higher-level CIO has the responsibility of overseeing and/or coordinating lower-level CIOs.
COO	Chief Operation Officer
C-Suite, C-Level	A general term to include all executives with a title containing the word "chief" or equivalent.
CTO	Chief Technology Officer
GS	General Schedule. The General Schedule is a position classification in the civil service of the US federal government, somewhat analogous to the ranks of colonel or below in the US Armed Forces. GS-15 is comparable to the rank of colonel in the Army and captain in the Navy; GS-7, to that of second lieutenant in the Army and ensign in the Navy.
HR	Human Resources
R&D	Research and Development
ROI	Return on Investment
SES	Senior Executive Service. The Senior Executive Service is a position classification in the civil service of the US federal government, somewhat analogous to the ranks of general or admiral in the US Armed Forces.
SVP	Senior Vice President
VC	Venture Capital; Venture Capitalist
VP	Vice President

Fast-Tracking Your Career: Soft Skills for Engineering and IT Professionals, First Edition.
Wushow "Bill" Chou.
© 2013 The Institute of Electrical and Electronics Engineers, Inc. Published 2013 by John Wiley & Sons, Inc.

Index

Fast-Tracking Your Career: Soft Skills for Engineering and IT Professionals, First Edition.
Wushow "Bill" Chou.
© 2013 The Institute of Electrical and Electronics Engineers, Inc. Published 2013 by John Wiley & Sons, Inc.